Post-translational Modification of Proteins by Lipids

A Laboratory Manual

Edited by

Urs Brodbeck · Clément Bordier

With 17 Figures

Springer-Verlag
Berlin Heidelberg New York
London Paris Tokyo

Prof. Dr. Urs Brodbeck
Institut für Biochemie
und Molekularbiologie
Bühlstraße 28
CH-3012 Bern, Switzerland

Dr. Clement Bordier
Biokema SA
2 ch. de la Chatanerie
CH-1023 Crissier-Lausanne, Switzerland

QH
450
.6
.P67
1988

ISBN 3-540-50215-7 Springer-Verlag Berlin Heidelberg New York
ISBN 0-387-50215-7 Springer-Verlag New York Berlin Heidelberg

Library of Congress Cataloging-in-Publication Data. Post-translational modification of proteins by lipids : a laboratory manuals / edited by Urs Brodbeck, Clement Bordier. p. cm.
Includes index. 1. Proteins–Post–translational modification–Laboratory manuals. 2. Lipids–
Physiological effect–Laboratory manuals. I. Brodbeck, U. (Urs), 1938-. II. Bordier, Clement.
[DNLM: 1. Lipids–laboratory manuals. 2. Membrane Proteins–biosynthesis–laboratory
manuals. 3. Translation, Genetic–laboratory manuals. QU 25 P857] QH450.6.P67 1988
574.19'245–dc 19 DNLM/DLC 88-29440

© Springer-Verlag Berlin Heidelberg 1988
Printed in Germany

Printing and binding: Druckhaus Beltz, Hemsbach/Bergstr.
2131/3130-543210 – Printed on acid-free paper

Preface

The growing interest in recent years in the anchoring to membranes of proteins by post-translational modification is documented by the large number of publications which appeared in this field. In September 1987, scientists from 10 countries from all over the world met in the resort village of Les Diablerets, Switzerland, to discuss the most recent advances made in this field. The sessions were devoted to the anchoring of membrane proteins by covalent attachment of fatty acids and of glycophospholipids. The workshop brought together many scientists working on vastly different proteins such as the variant surface glycoprotein of Trypanosomes and antigens of the mammalian cells. The subject of the workshop unified many scientists who had not met before and thus greatly stimulated interdisciplinary work.

In addition to the lectures, each participant was provided with a collection of **Methods** currently in use in the study of membrane proteins anchored by post-translational modification. An updated version of this collection is now presented as a **Laboratory Manual**, and the techniques described therein will give researchers easy and practical access to the investigation of post-translationally modified proteins. The publication of the present book by **Springer** follows an established tradition of previously published manuals on the handling of membrane proteins.

Our thanks go to the authors who made the essential contribution in writing and adapting the experimental protocols, to Mrs. R. Schneider for her diligence and effort in preparing the camera-ready version of this book, to Dr. S. Stieger for her careful proofreading, and to Mrs. B. Brodbeck for help in creating the titlepage.

Bern/Lausanne, July 1988

URS BRODBECK
CLEMENT BORDIER

Contents

Contributors

You will find the addresses at the beginning of the respective contribution.

Berger, M.   94
Bon, S.   132
Bordier, C.   29
Brodbeck, U.   34
Burns, G.R.   94
Buss, J.E.   64
Callahan, F.E.   82
Capdeville, Y.   43
Casey, P.J.   64
Chettibi, S.   123
Davitz, M.A.   40
Deregnaucourt, C.   43
Edelman, M.   82
Englund, P.T.   9
Ferguson, M.A.J.   1
Haas, R.   22
Hart, G.W.   9
Hayashi, S.   88
Hereld, D.   9
Ikehara, Y.   16
Ikezawa, H.   127
Koch, N.   119
Krakow, J.L.   9

Lambrecht, B.   46
Lawrence, A.J.   123
Lehmann, L.   76
Magee, A.I.   59
Massoulié, J.   132
Mattoo, A.K.   82
McDowell, W.   99
Nayudu, P.R.V.   46
Norman, H.A.   82
Puklavec, M.   51
Rosenberry, T.L.   22
St.John, J.B.   82
Schmidt, M.   94
Schmidt, M.F.G.   46, 94
Schwarz, R.T.   99
Sigel, E.   76
Staufenbiel, M.   72
Stieger, S.   34
Stucki, J.W.   76
Taguchi, R.   127
Takami, N.   16
Toutant, J.P.   132
Williams, A.F.   51
Wu, H.C.   88

.

Identification of Glycosyl-Phosphatidylinositol Membrane Anchors by Fatty Acid Labeling

M.A.J. Ferguson

Department of Biochemistry, University of Dundee, Dundee, Scotland, U.K.

INTRODUCTION

Glycosyl-phosphatidylinositol (G-PI) membrane anchors are employed by a number of eukaryotic plasma membrane glycoproteins. Examples include protozoal antigens e.g., the variant surface glycoprotein (VSG) of *Trypanosoma brucei*; enzymes e.g., alkaline phosphatase and acetylcholinesterase; T-cell components e.g., Thy-1 and T-cell activating protein; complement regulating proteins e.g., decay accelerating factor; and cell-adhesion molecules e.g., N-CAM$_{120}$. The occurance and biochemistry of G-PI anchored proteins has been reviewed in detail (Low et al. 1986; Cross 1987; Low 1987).

The physiological significance of using a G-PI anchor in place of a conventional transmembrane polypeptide domain is still unclear, however, possible advantages are:

a) a higher degree of lateral mobility in the membrane bilayer (Woda and Gilman 1983; Ishihara et al. 1987),

b) conservation of space in the membrane, and

c) the potential to release proteins from membranes by stimulating endogenous G-PI-specific phospholipase C (Bulow and Overath 1986; Fox et al. 1986; Hereld et al. 1986).

The best characterized G-PI anchor is that of *T. brucei* VSG (Ferguson et al. 1985a; Ferguson et al. 1985b). The complete structure of this G-PI anchor has recently been completed (Ferguson et al. 1988) and is shown schematically in Figure 1.

IDENTIFICATION OF G-PI ANCHORS

PI-PLC release

Many G-PI anchors have been identified by the use of bacterial phosphatidylinositol-specific phospholipase C (PI-PLC) from either *Staphylococcus aureus* or *Bacillus thuringiensis* (Low

1987). In this case, the PI-PLC cleaves the diacyl-glycerol moiety from the anchor and thereby liberates the attached protein from the membrane. The release may be monitored by immuno-precipitation of the supernatant or by FACS analysis of cells before and after PI-PLC treatment using an antibody directed against the candidate protein (Low and Kincade 1985; Davitz et al. 1986; Koch et al. 1986).

This approach is limited by the need for a probe or assay for the candidate protein and the fact that some G-PI anchors are PI-PLC resistant (Low 1987).

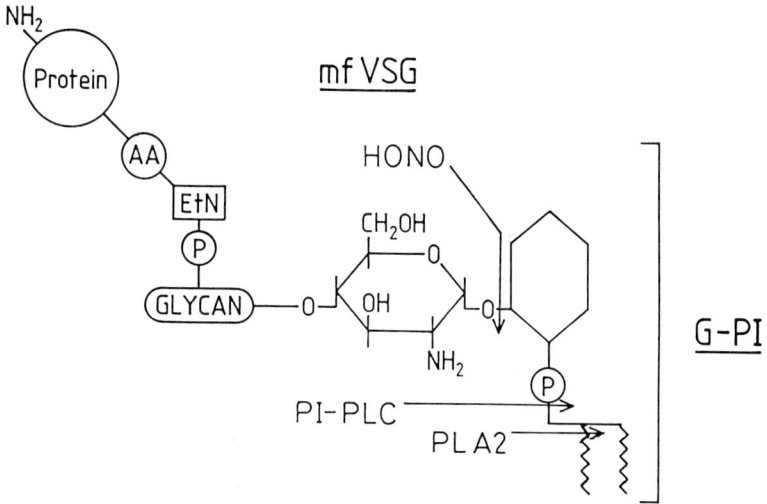

Fig. 1

CRD blotting

Rabbit antisera raised against soluble form VSGs (sVSG) which lacks the diacyl-glycerol moiety of the G-PI anchor, contain a population of antibodies directed against an epitope in the carbohydrate of the G-PI. This fraction of antibodies will cross-react with all VSGs exclusively in this cross-racting determinant or CRD (Holder 1985). Anti-CRD antibody has been used to detect the CRD epitope in the soluble forms of Leishmania GP-63 protease, Torpedo and human erythrocyte acetylcholinesterase, and human DAF and alkaline phosphatase (Bordier et al. 1986; Stieger et al. 1986; Davitz et al. 1987).

This approach is limited by the need to generate the soluble form of the protein by PI-PLC action. The CRD epitope is cryptic in the intact membrane form of the proteins.

Biosynthetic labeling

Labeling with G-PI-specific precursors such as [^{3}H]ethanolamine and [^{3}H]*myo*-inositol strongly suggest the presence of a G-PI anchor (reviewed in Low 1987). Fatty acid labeling is initially more ambiguous as other types of protein acylation are common. However, subsequent enzymatic and chemical treatments of [^{3}H]fatty acid labeled proteins will quickly provide quite detailed information on the nature of the fatty acid linkage and will give unambiguous information on the presence or absence of a G-PI anchor.

These experiments assume that a protein of interest has been labeled with [^{3}H]fatty acid (Ferguson and Cross 1984) e.g., [^{3}H]myristate or [^{3}H]palmitate, and purified by RP-HPLC (Clarke et al. 1985), preparative SDS-PAGE, and electroelution (Haldar et al. 1985), or other means such as affinity chromatography. The examples given are for [^{3}H]myristic acid labeled membrane form VSG (mfVSG), but any [^{3}H]fatty acid labeled protein can be used.

EXPERIMENTAL PROCEDURES

Base hydrolysis

The first question to ask is whether or not the fatty acid is base-stable (and, therefore, probably amide linked to the N-terminus of the protein) or base-labile, in which case ester linkages are involved (Ferguson and Cross 1984).

Suspend 10,000 cpm aliquots of [^{3}H]mfVSG in 200 μl 90% ethanol (control) and 200 μl 90% ethanol containing 50 mM NaOH. After 40 min at room temperature, add 800 μl of 10% (v/v) acetic acid to acidify the mixture and vortex with 1 ml of toluene. Centrifuge to separate the phases and count 0.4 ml of the upper toluene phase in 10 ml of scintillation fluid. Record the % of [^{3}H]fatty acid released by base hydrolysis:

$$\% \text{ released} = \frac{(+\text{NaOH counts}) - (-\text{NaOH counts})}{10,000 - (-\text{NaOH counts})} \times 2.5 \times 100$$

To check the nature of the released product (i.e. free-fatty acid), dry a further 0.5 ml of the toluene phase, redissolve in about 10 μl of "Standard A1" and apply to a TLC-plate (System A).

If all or most of the [^{3}H]fatty acid is released under these conditions as free fatty acid, then an ester linkage can be assumed. This could mean either ester linkage to serine/threonine, thioester linkage to cysteine, or ester linkage to glycerol in a G-PI anchor. The next experiments are specific for G-PI anchor identification.

Acetolysis

Acetolysis may remove [3H]fatty acid from other acylated proteins as free fatty acid, but acetolysis of G-PI anchors will produce a diacylglycerol acetate which can be identified by TLC.

a) Dry 10,000 cpm of [3H]mfVSG in a reaction vial,

b) add 200 μl acetic acid - acetic anhydride (3:2) and place in a heating block at 105°C for 3 h.

c) Dry the reaction mixture under a stream of N_2, add 160 μl $CHCl_3$, 80 μl MeOH, and 60 μl H_2O and vortex.

d) Recover the lower chloroform rich phase with a pasteur pipette (with care!) and transfer to a preweighed (weight A) 1 ml screw cap vial (use a 4 or 5 place balance). Re-extract the upper aqueous phase with 200 μl of pre-equilibrated lower phase. Transfer the lower phase to the same 1 ml vial. Weigh the vial plus pooled lower phases (weight B), then transfer a 40 μl sample to a scintillation vial. Allow the chloroform to evaporate, then add scintillation fluid and count. Weigh the vial again (weight C). Calculate the % of [3H]-fatty acid released from the protein as follows:

$$\% \text{ released } = \frac{(B - A) \text{ x cpm counted}}{(B - C) \text{ x } 10,000} \text{ x } 100$$

Take the remainder of the sample in the 1 ml vial and dry it under a stream of N_2, redissolve it in 10 μl of "Standard A2" and apply to a TLC-plate for product analysis (System A).

If all or most of the [3H]fatty acid is released as diglyceride acetate under these conditions, then it can be assumed that the fatty acids were originally present as diglyceride, most likely as part of a phospholipid structure (Ferguson et al. 1985a).

PI-PLC treatment

Resuspend two 10,000 cpm aliquots of [3H]mfVSG in 500 μl 25 mM HEPES-NaOH buffer, pH 7.4, containing 0.1% (w/v) sodium deoxycholate. To one sample, add 5 μl of PI-PLC stock solution (about 0.5 μg of enzyme) and incubate at 37°C for 1 h. Add 1 ml of toluene, vortex, and centrifuge to separate the phases. Take 0.4 ml of the upper toluene phase and add it to 10 ml of scintillation fluid in a scintillation vial and count. Calculate the % [3H]fatty acid released as follows:

$$\% \text{ released} = \frac{\text{(+PI-PLC counts)} - \text{(-PI-PLC counts)}}{10,000 - \text{(-PI-PLC counts)}} \times 2.5 \times 100$$

Take a further 500 μl of the +PI-PLC toluene phase, dry, redissolve in 10 μl "Standard A1", and analyse the nature of the products by TLC (System A). If all or most of the [^{3}H]-fatty acid is released as diglyceride, it can be assumed that the fatty acid is present as a phosphoinositide (i.e. G-PI).

PLA$_2$ treatment

Resuspend two 10,000 cpm aliquots of [^{3}H]mfVSG in 0.5 ml of 0.5 M Tris-HCl, pH 7.5, 15 mM CaCl$_2$ containing 0.05% NP-40. To one, add 50 μl of 1 mg/ml PLA$_2$ solution and incubate for 1 h at 37°C. Acidify with 50 μl CHCl$_3$ and add 1 ml toluene, vortex, and centrifuge. Count 0.4 ml of the upper toluene phase. Calculate the % [^{3}H]fatty acid released as follows:

$$\% \text{ released} = \frac{\text{(+PLA}_2 \text{ counts)} - \text{(-PLA}_2 \text{ counts)}}{10,000 - \text{(-PLA}_2 \text{ counts)}} \times 2.5 \times 100$$

Take a further 0.5 ml of the PLA$_2$ toluene phase, dry, redissolve in "Standard A1", and analyse the products by TLC (System A). If about 50% of the [^{3}H]fatty acids are released by PLA$_2$ as free fatty acid, then a phospholipid structure may be assumed. This enzyme digest may work even when PI-PLC does not.

Nitrous acid deamination

Resuspend two 10,000 cpm aliquots of [^{3}H]mfVSG in 75 μl of 100 mM sodium acetate/acetic acid buffer, pH 4.0. To one, add 75 μl of freshly prepared 0.5 M NaNO$_2$ (sodium nitrite), and to the other 75 μl of 0.5 M NaCl. Incubate at room temperature for 3 h. Add 15 μl of 1 M HCL, 400 μl of CHCl$_3$, and 200 μl MeOH. Vortex and separate the phases by centrifugation. Recover the lower CHCl$_3$ phase with a pasteur pipette and transfer it to a preweighed 3.5 ml glass vial (weight A). Re-extract the aqueous phase with 0.5 ml of pre-equilibrated lower phase. Weigh the vial plus the pooled lower phases (weight B). Take a 200 μl sample and dry it in a scintillation vial for counting. Reweigh the sample vial (weight C). Calculate the % [^{3}H]fatty acid released as follows:

$$\text{\% released} = \frac{(+\text{NaNO}_2 \text{ counts}) - (-\text{NaNO}_2 \text{ counts})}{10,000 - (-\text{NaNO}_2 \text{ counts})} \times \frac{(B - A)}{(B - A)} \times 100$$

Dry the rest of the pooled lower phases, redissolve the products in 10 μl "Standard B1", and apply to a TLC-plate (System B). If most of the [^{3}H]fatty acid is released as phosphatidylinositol, it can be assumed that a G-PI anchor is present where the PI-group is substituted by a non-N-acetylated hexosamine residue (Ferguson et al. 1985b).

TLC SYSTEMS

System A

Using Silica gel G or Si 60 TLC-plates activated at 125°C for 1 h before use.

Solvent:

Petroleum ether - diethylether - acetic acid (80:20:1)

Standard A1:

Use 10 μl of 10 mg/ml (each in chloroform):

monomyristin

1,2 dimyristin

1,3 dimyristin

myristic acid

trimyristin

methylmyristate

Standard A2:

Use 10 μl of 10 mg/ml (each in chloroform):

monomyristin-diacetate

1,2 dimyristin-acetate

1,3 dimyristin-acetate

Prepared by acetylating monomyristin, 1,2 and 1,3 dimyristin in acetic anhydride - pyridine (1:1) for 12 h at room temperature. Dry the products under a stream of N_2 and redissolve in $CHCl_3$.

System B

Using Si 60 TLC-plates activated at 125°C for 1 h before use.

Solvent:

Develop once with acetone - petroleum ether (1:3), dry, then develop with $CHCl_3$ - MeOH - H_2O - acetic acid (25:15:4:2).

Standard B:

Use 20 μl of 10 mg/ml (each in chloroform - methanol (2:1):

phosphatidic acid

phosphatidylcholine

phosphatidylethanolamine

phosphatidylglycerol

phosphatidylinositol

Detection

Visualize the standards with I_2 vapour. Detect the ^{3}H-labeled products with a TLC-scanner (linear analyser), or by scraping 0.5 cm strips into scintillation fluid.

REFERENCES

Bordier C, Etges RJ, Ward J, Turner MJ, Cardoso de Almeida ML (1986) Proc Natl Acad Sci USA 83:5988-5991

Bulow R, Overath P (1986) J Biol Chem 261:11918-11923

Clarke MW, Olafson RW, Pearson TW (1985) Mol Biochem Parasitol 17:19-34

Cross GAM (1987) Cell 48:179-181

Davitz MA, Low MG, Nussenzweig V (1986) J Exp Med 163:1150-1161

Davitz MA, Gurnett AM, Low MG, Turner MJ, Nussenzweig V (1987) J Immunol 138:520-523

Ferguson MAJ, Cross GAM (1984) J Biol Chem 259:3011-3015

Ferguson MAJ, Haldar K, Cross GAM (1985a) J Biol Chem 260:4963-4968

Ferguson MAJ, Low MG, Cross GAM (1985b) J Biol Chem 260:14547-14555

Ferguson MAJ, Homans SW, Dwek RA, Rademacher TW (1988) Science 239:753-759

Fox JA, Duszenko M, Ferguson MAJ, Low MG, Cross GAM (1986) J Biol Chem 261:15167-15171

Haldar K, Ferguson MAJ, Cross GAM (1985) J Biol Chem 260:4963-4968

Hereld D, Krakow JL, Bangs JD, Hart GW, Englund PT (1986) J Biol Chem 261:13813-13819

Holder AA (1985) Curr Top Microbiol Immunol 117:57-74

Ishihara A, Hou Y, Jacobson K (1987) Proc Natl Acad Sci USA 84:1290-1293

Koch F, Thiele HG, Low MG (1986) J Exp Med 164:1338-1343

Low MG, Kincade PW (1985) Nature 318:62-64

Low MG, Ferguson MAJ, Futerman AH, Silman I (1986) TIBS 11:212-215
Low MG (1987) Biochem J 244:1-13
Stieger A, Cardoso de Almeida ML, Blatter MC, Brodbeck U, Bordier C (1986) FEBS Lett 199: 182-186
Woda BA, Gilman SC (1983) Cell Biol Int Rep 7:2037-2039

[³H]Myristate-Labeled Variant Surface Glycoprotein from Trypanosoma brucei: Preparation and Use in the Assay of Glycan-Phosphatidylinositol-Specific Lipases

D. Hereld, J.L. Krakow, G.W. Hart, and P.T. Englund

Department of Biological Chemistry, The Johns Hopkins University School of Medicine, 725 N. Wolfe Street, Baltimore, Maryland 21205, USA

INTRODUCTION

A rapidly growing number of membrane proteins have been shown to possess structurally related, covalently attached glycolipid membrane anchors (Cross 1987; Low 1987). These proteins are functionally diverse and are found in organisms as unrelated as the protozoan trypanosomes and man. The glycolipid anchor of the variant surface glycoprotein (VSG) of *Trypanosoma brucei* has been studied in the greatest detail. Certainly one reason for this is the ease with which hundreds of milligrams of pure VSG can be obtained.

Trypanosomes have also been shown to possess a phospholipase C which cleaves the membrane-form of VSG (mfVSG) to yield soluble VSG (sVSG) and dimyristoylglycerol (Cardoso de Almeida and Turner 1983; Ferguson et al. 1985). This enzyme which we designate VSG lipase, has been purified to homogeneity and shown to be specific for the glycolipid anchor of mfVSG (Bulow and Overath 1986; Fox et al. 1986; Hereld et al. 1986). It cleaves the phosphatidylinositol moiety of the anchor in preference to free phosphatidylinositol.

The ability to radiolabel mfVSG with [³H]myristic acid (Ferguson and Cross 1984) has greatly facilitated the study of VSG lipase. [³H]Myristate-labeled mfVSG which we will abbreviate as [³H]mfVSG, serves as the substrate in a rapid, sensitive, and quantitative assay for VSG lipase. Furthermore, [³H]mfVSG has been instrumental in the recent discovery and study of two other glycan-phosphatidylinositol-specific lipases. One of these, a phospholipase C like VSG lipase, has been purified from rat hepatocyte plasma membranes (Fox et al. 1987). The other is a human serum phospholipase D which liberates phosphatidic acid from mfVSG (Davitz et al. 1987).

Here, we describe procedures for the preparation of [³H]mfVSG and its use as substrate in a simple and quantitative assay for VSG lipase.

EXPERIMENTAL PROCEDURES

A. Growth and isolation of trypanosomes

1. Inject 10^6-10^7 bloodstream-form trypanosomes (we use *Trypanosoma brucei* variant ILTat 1.3) into each of 4 adult Swiss mice intraperitoneally.

2. Exsanguinate the mice when the parasite density is around 10^9/ml of blood (usually 2 or 3 days after infection). This can be assessed by microscopic examination of tail blood; there will be nearly equal numbers of trypanosomes and erythrocytes at this density. Without intervention, mice usually die within several hours at this stage of an infection. Bleed each mouse as follows:

   a) Etherize a mouse just until it stops breathing.

   b) With scissors and forceps, cut through the sternum lengthwise to open the chest cavity.

   c) Using a 1 ml syringe, inject 0.1 ml of BBS (50 mM NaCl, 5 mM KCl, 55 mM D-glucose, 50 mM Bicine, adjusted to pH 8.0 with NaOH) containing heparin (100 units/ml) into the right ventricle of the heart.

   d) Without removing the syringe, withdraw as much blood from the heart as possible. As much as 1 ml can be obtained from a full-grown mouse. Remove the needle from the syringe and transfer the blood to a tube containing 2 ml of BBS containing 10 units of heparin/ml (BBSH) on ice.

   e) To collect additional blood, cut the heart and both the aorta and inferior vena cava just above the diaphragm. Add a few drops of BBSH to the blood which pools in the chest cavity, and combine it with the other blood.

3. At 4°C, apply the blood from 4 mice to a 20 ml column of DEAE cellulose (Whatman, type DE52) equilibrated with BBS. A 50 ml syringe barrel plugged with cotton works well for this. When the blood has run into the column bed, wash the column with BBS. Under these conditions, mouse blood cells bind to the column and trypanosomes flow through (Lanham and Godfrey 1970).

4. Begin collecting the column effluent in a 50 ml plastic culture tube on ice when the trypanosomes first emerge from the column in pearly white drops and continue until about 30 ml have been collected.

5. Centrifuge the cells (Sorvall HS-4 rotor) at 2000 rpm at 4°C for 5 min, and resuspend them in 16 ml ice-cold BBS.

6. Using a haemacytometer, determine the cell density; a 200-fold dilution in BBS is usually appropriate for this measurement.

7. Transfer 2×10^9 cells to a fresh 50 ml tube. Wash the cells in 10 ml ice-cold BBS and centrifuge again.

B. Labeling trypanosomes with [³H]myristate

(Essentially the method of Ferguson and Cross 1984)

1. In advance, prepare labeling medium as follows:

   a) Place 1 mCi of [9,10-³H]myristic acid (Amersham, 55 Ci/mmol, in toluene) in a 1.5 ml microfuge tube and evaporate to dryness in a Speed-Vac concentrator (Savant).

   b) Dissolve the fatty acid in 20 μl of 95% ethanol.

   c) Form a BSA-[³H]myristate complex by adding 0.25 ml of 20 mg/ml defatted BSA (Sigma, cat. # A-6003) in 150 mM NaCl, 10 mM NaP$_i$, pH 7.0. Mix thoroughly.

   d) Combine this mixture with 10 ml of RPMI-1640 culture medium (Gibco) supplemented with 0.5 mg/ml defatted BSA and 25 mM Hepes buffer, pH 7.4.

2. Resuspend the washed cells (from Section A) in 10 ml ice-cold labeling medium and incubate the suspension at 37°C for 60-90 min.

3. Swirl the suspension periodically, and examine some cells under the microscope; they should remain slender and motile.

4. If the medium becomes acidic (yellow), titrate it back to near pH 7.4 (orange) by adding small volumes of 1 N NaOH while mixing.

5. After labeling, shift the suspension to ice and let it cool for 5-10 min. Centrifuge the cells as before.

6. Wash the labeled cells with 10 ml cold BBS and centrifuge again.

7. The medium recovered in step 5 can be stored at -20°C and reused at least once more. Before reusing it, we adjust the pH to about 7.4 with 1 N NaOH and add 1 mg of D-glucose/ml of medium.

C. Isolation of [³H]myristate-labeled mfVSG

mfVSG is stably anchored to intact trypanosomes. However, when they are lysed under non-denaturing conditions and warmed to 37°C, mfVSG is rapidly cleaved by the endogenous VSG lipase. Therefore, it is necessary to inactivate the lipase in order to isolate [³H]mfVSG. This is accomplished by lysing the cells in the presence of p-chloromercuriphenylsulfonate (pCMPS) which inhibits VSG lipase (Turner et al. 1985). [³H]mfVSG isolated in this way, contains [³H]myristate-labeled lipids which would interfere in the assay for VSG lipase (and other glycan-phosphatidylinositol-specific lipases) described in Section D. These interfering lipids are readily removed by extraction with n-butanol. Figure 1 shows representative fractions from the isolation of [³H]mfVSG.

The procedure is as follows:

1. In advance, prepare 100 mM pCMPS by dissolving 42 mg of sodium pCMPS (Sigma) in

1 ml of 0.1 N NaOH. Store the solution at 4°C.

2. Lyse the labeled cells (from Section B) by resuspending them in 4 ml ice-cold 10 mM sodium phosphate, 1 μg/ml leupeptin, 0.1 mM tosyl-lysine-chloromethylketone, pH 7.0 (Hypotonic Lysis Buffer). Transfer the suspension to a 30 ml Corex tube on ice and incubate for 5 min. Very little mfVSG is cleaved by VSG lipase during this brief period on ice.

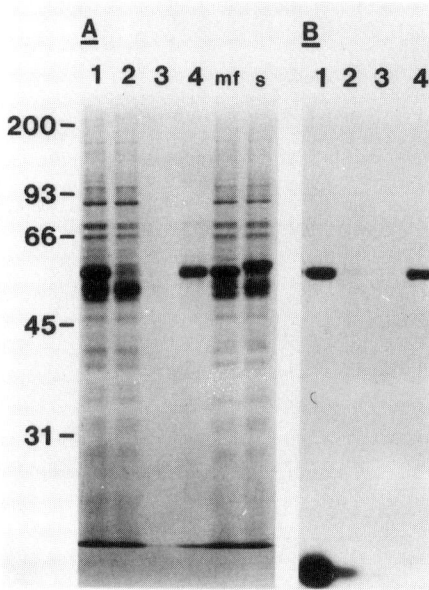

Fig. 1. Purification of [^{3}H]myristate-labeled mfVSG. Various fractions from the purification of [^{3}H]myristate-labeled mfVSG were analyzed by SDS-polyacrylamide gel elctrophoresis. The gel was stained with Coomassie blue (**A**), and then fluorographed (**B**). The fractions depicted in both panels and the corresponding steps from "Experimental Procedures" (in parenthesis) are as follows: lane 1, whole cells (step C.2); lane 2, soluble fraction from pCMPS lysis (step C.3); lane 3, supernatant from wash of pCMPS insoluble fraction (step C.4); lane 4, pCMPS insoluble fraction after n-butanol extraction (step C.11). Lanes "mf" and "s" indicate the positions of mfVSG and sVSG, respectively. VSG is the prominant band in each. The positions and molecular weights (expressed in kDa) of marker proteins are shown to the left of panel A.

3. Add 0.21 ml of ice-cold 100 mM pCMPS. After 5 min on ice, centrifuge the suspension (Sorvall HB-4 rotor) at 6500 rpm at 4°C for 15 min.

4. Wash the pellet with 4 ml of Hypotonic Lysis Buffer plus 0.21 ml of 100 mM pCMPS and centrifuge again.

5. Dissolve the washed pellet in 2 ml of 1% SDS by heating in a water bath at 100°C and vortex.

6. At room temperature, add 20 ml of water-saturated n-butanol to the SDS solution. Mix thoroughly, and then resolve the aqueous and organic phases by centrifugation (Sorvall HB-4 rotor) at 8500 rpm for 20 min.

7. Remove as much of the upper, organic phase as possible leaving behind the lower, aqueous phase and an opaque intermediate phase. The latter will be diffuse initially but contracts with repeated extraction.

8. Replace the upper phase with an equal volume of water-saturated n-butanol. Repeat the extraction several times until the total ^{3}H radioactivity in the final extract is 2×10^5 dpm or less.

9. Eliminate the aqueous phase by performing a final extraction with n-butanol (not water-saturated).

10. Recover the gummy white precipitate by centrifugation, and wash it with several milliliters of anhydrous ether to remove n-butanol. Allow residual ether to evaporate.

11. Dissolve the pellet in 0.5 ml of 1% SDS as before. Remove insoluble debris by centrifuging in a microfuge at room temperature for 5 min. Transfer the supernatant to a clean tube.

12. Determine the protein concentration of the preparation by the method of Lowry et al. 1951). Expect about 4 mg/ml.

13. The [^{3}H]mfVSG obtained is essentially pure in terms of protein (Figure 1A, lane 4) and ^{3}H radioactivity (Figure 1B, lane 4). Its specific radioactivity is usually between 6 and 12 dpm/ng of protein. All of the [^{3}H]mfVSG can be cleaved by VSG lipase, and it is stable at -70°C for a year or more.

D. Assay for VSG lipase

VSG lipase can be assayed by its ability to cleave [^{3}H]mfVSG generating [^{3}H]dimyristoyl-glycerol and unlabeled sVSG. [^{3}H]Dimyristoylglycerol, unlike [^{3}H]mfVSG, is extractable by n-butanol; thus, the amount of labeled product generated can be simply and sensitively measured. Some properties of the VSG lipase assay are shown in Figure 2.

The assay is performed as follows:

1. Dilute the [^{3}H]mfVSG (from Section C) to 2 μg/15 μl in 1% NP40, 50 mM Tris-HCl, 5 mM EDTA, pH 8.0 (assay buffer).

2. In a 1.5 ml microfuge tube on ice combine 15 μl [^{3}H]mfVSG (2 μg) in assay buffer and 10 μl VSG lipase diluted in assay buffer.

3. Incubate the mixture at 37°C for 30 min, and return it to ice.

4. Add 0.5 ml water-saturated n-butanol (room temperature) and vortex vigorously.

5. Centrifuge for 1 min to separate the organic and aqueous phases.

6. Measure the ^{3}H radioactivity in 0.4 ml (80%) of the upper, organic phase by scintillation counting using a scintillation fluid for aqueous samples.

7. Simultaneously, determine background radioactivity in a blank without enzyme.

8. In the VSG lipase titration shown in Figure 2A, the assay behaves linearly in the range of 0 to 0.5 μg of [^{3}H]mfVSG. For quantitative determinations, we use only data in this range.

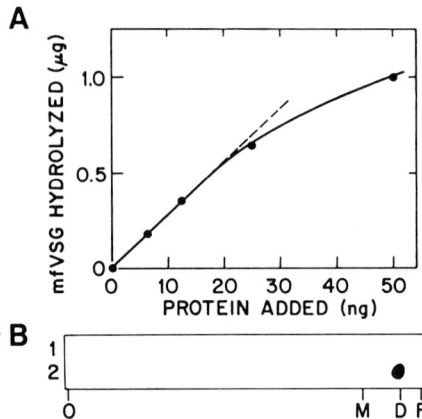

Fig. 2. VSG lipase assay. (A) The assay for VSG lipase was performed as described in the text with varying amounts of a partially purified preparation of VSG lipase. The amount of mf-VSG hydrolyzed was calculated from its specific radioactivity (11.0 dpm/ng). Data points represent averages of duplicate determinations which differed by 5% or less. (B) Products from assay reactions without (lane 1) or with (lane 2) VSG lipase were analyzed by silica gel thin layer chromatography followed by fluorography. The single labeled product of VSG lipase treatment comigrated with a dimyristoylglycerol marker (D). The positions of the sample origin (O), solvent front (F), and myristic acid (M) are also indicated.

COMMENTS

The assay conditions for VSG lipase described here may require modification when other glycan-phosphatidylinositol-specific lipases capable of cleaving [^{3}H]mfVSG are assayed by this method. For instance, Ca^{2+} may have to be added to the assay buffer for lipases such as the human serum phospholipase D (Davitz et al. 1987) which require this divalent cation. The pH, ionic strength, detergent concentration, and duration and temperature of the reaction may also have to be optimized for other lipases.

We believe that [³H]mfVSG will prove to be a suitable substrate for other glycan-phosphatidylinositol-specific lipases from a variety of sources. Therefore, we hope that these methods will be useful to investigators studying enzymes which specifically cleave glycolipid anchors on proteins.

REFERENCES

Bulow R, Overath P (1986) J Biol Chem 261:11918-11923
Cardoso de Almeida ML, Turner MJ (1983) Nature 302:349-352
Cross GAM (1987) Cell 48:179-181
Davitz MA, Hereld D, Shak S, Krakow J, Englund PT, Nussenzweig V (1987) Science 238: 81-84
Ferguson MAJ, Cross GAM (1984) J Biol Chem 259:3011-3015
Ferguson MAJ, Haldar K, Cross GAM (1985) J Biol Chem 260:4963-4968
Fox JA, Duszenko M, Ferguson MAJ, Low MJ, Cross GAM (1986) J Biol Chem 261:15767-15771
Fox JA, Soliz NM, Saltiel AR (1987) Proc Natl Acad Sci USA 84:2663-2667
Hereld D, Krakow JL, Bangs JD, Hart GW, Englund PT (1986) J Biol Chem 261:13813-13819
Lanham SM, Godfrey DG (1970) Exp Parasitol 28:521-534
Low MG (1987) Biochem J 244:1-13
Lowry OH, Rosebrough NJ, Farr AL, Randall RJ (1951) J Biol Chem 193:265-275
Turner MJ, Cardoso de Almeida ML, Gurnett AM, Raper J, Ward J (1985) Curr Top Microbiol Immunol 117:23-55

Identification of the Glycolipid Anchor of Alkaline Phosphatase by Metabolic Labeling

Y. Ikehara* and N. Takami[+]

*Department of Biochemistry, and [+]Radioisotope Laboratory, Fukuoka University, School of Medicine, 45-1,7-chome Nanakuma, Jonan-ku, Fukuoka 814-01, Japan

INTRODUCTION

An increasing number of membrane proteins have been shown to possess covalently attached glycolipid membrane anchors (Low et al. 1986; Cross 1987). The glycolipid anchors of the variant surface glycoprotein of *Trypanosoma brucei* (Ferguson et al. 1985) and Thy-1 (Tse et al. 1985) have been well characterized by chemical analysis demonstrating that the anchor is the 1,2-diacylglycerol moiety of a phosphatidylinositol moiety which is covalently linked to the polypeptide chains *via* other components including ethanolamine and carbohydrates.

The documentation of such a unique anchoring has primarily relied on studies with phosphatidylinositol-specific phospholipase C (PI-PLC) and on chemical analysis. Another useful method for identification of the anchor is characterization of proteins which are metabolically labeled with radioactive compounds to be incorporated into the glycolipid. Proteins labeled with such compounds could be easily identified by SDS-polyacrylamide gel electrophoresis/fluorography (Ferguson and Cross 1984; Fatemi and Tartakoff 1986; Takami et al. 1988), and some of the incorporated glycolipid components could be specifically removed by PI-PLC or nitrous acid deamination, while the proteins labeled with radioactive amino acids remain unchanged (Takami et al. 1988).

Alkaline phosphatase (ALP) is the first example of proteins which have been demonstrated to be released from membranes by PI-PLC (Ikezawa et al. 1976; Low and Finean 1977; Kominami et al. 1985). Most recently, we have demonstrated the presence of the glycolipid anchor in human placental ALP by metabolic labeling experiments (Takami et al. 1988). The use of human choriocarcinoma cells, JEG-3 (Kohler and Bridson 1971), was of great advantage to our experimental purpose because the cells produce placental ALP which is markedly induced by treatment with sodium butyrate (Ito and Chou 1984).

MATERIALS

1. Cells

Human choriocarcinoma cells, JEG-3 (Kohler and Bridson 1971) provided by Dr. S. Sekiya (Chiba University School of Medicine, Chiba, Japan).

2. Culture medium

RPMI-1640 medium (Nissui Seiyaku, Tokyo) supplemented with penicillin (100 units/ml), streptomycin (100 μg/ml), gentamycin (50 μg/ml), and fetal calf serum (10%). Eagle's minimum essential medium (Nissui Seiyaku) was also used for some labeling experiments.

3. Labeled compounds

L-[^{35}S]methionine (1,120 Ci/mmol), [9,10-^{3}H]palmitic acid (30.0 Ci/mmol), [9,10-^{3}H]stearic acid (21.5 Ci/mmol, prepared by special order), myo-[2-^{3}H]inositol (17.1 Ci/mmol), [2-^{3}H]mannose (28.6 Ci/mmol), [6-^{3}H]galactose (29.7 Ci/mmol), and [1,6- ^{3}H]glucosamine (39.3 Ci/mmol) were obtained from New England Nuclear (Boston, MA); [1-^{3}H]ethanolamine (12.0 Ci/mmol) from Amersham (Arlington Height, IL).

4. Pansorbin (fixed *Staphylococcus aureus* cells)

Pansorbin (Calbiochem-Behring, La Jolla, CA) was incubated for 30 min with shaking at room temperature with 1% bovine serum albumin (BSA) in solution A (50 mM Tris-HCl, pH 7.4; 190 mM NaCl; 6 mM EDTA; 2.5% Triton X-100; 0.02% NaN$_3$, and Trasylol, 100 units/ml), washed twice with solution A, and resuspended in the same solution.

5. Antibodies

Placental ALP was purified by the same method as described previously (Miki et al. 1986a), and used as the antigen for production of anti-(placental ALP)antibodies in rabbits (Takami et al. 1988).

6. PI-PLC

The enzyme was purified from *Bacillus cereus* T as described previously (Kominami et al. 1985).

7. Other materials

Tunicamycin and Trasylol (Sigma Chemcials, St. Louis, MO), N-glycanase (Genzyme Corp., Boston, MA), and proteinase inhibitors (antipain, elastatinal, leupeptin, chymostatin, pepstatin A, and phosphoramidon) (Protein Research Foundation, Osaka, Japan).

EXPERIMENTAL PROCEDURES

1. Induction of ALP with sodium butyrate

JEG-3 cells were cultured at $37^{\circ}C$ in 60 mm Falcon dishes containing 3 ml of RPMI-1640 medium supplemented with antibiotics and foetal calf serum as described above. Before being used in the following experiments, cells (3×10^6 cells/dish) were pretreated for 1 day with 1.5 mM sodium butyrate for the induction of ALP (Ito and Chou 1984; Takami et al. 1988). Sodium butyrate was also present in the medium throughout preincubation and pulse-chase periods.

2. Metabolic labeling procedures

a) For labeling with $[^{35}S]$methionine, cells were preincubated at $37^{\circ}C$ for 1 h with RPMI-1640 medium lacking methionine and calf serum. The cells were labeled with $[^{35}S]$methionine (100 μCi/2.0 ml/dish) in the above medium for the given times, and then, if necessary, chased in the complete RPMI-1640 medium. Cells were also labeled with $[^3H]$ethanolamine (500 μCi/dish) in 2.0 ml of the complete RPMI-1640 medium.

b) For labeling with $[^3H]$palmitate or $[^3H]$stearate, appropriate amounts of the radioactive compounds in ethanol were evaporated to dryness under N_2 just before use, and dissolved again in small volumes of ethanol (2 or 5 mCi/10-15 μl). 2 mCi/10-15 μl of $[^3H]$palmitate or $[^3H]$stearate were added to the cell culture which had been preincubated for 1 h in 2 ml of RPMI-1640 medium lacking calf serum. After the medium was immediately mixed well, cells were incubated at 37°C for 4 to 14 h. When cells were labeled for 20 min, 5 mCi/10-15 μl of the ^{3}H-fatty acid was used.

c) When cells were labeled with [^{3}H]inositol, they were preincubated for 1 h in 2 ml of Eagle's minimum essential medium (MEM), and then labeled with [^{3}H]inositol (500 μCi/ dish) in 2.0 ml of the same medium at 37°C for 8 to 14 h.

d) For labeling carbohydrate moieties of ALP, cells were incubated for 8 h with ^{3}H labeled mannose, galactose, or glucosamine (200 μCi each/dish) in 2.0 ml of Eagle's MEM modi- fied to contain 10 mM pyruvate and 0.56 mM glucose (Miki et al. 1986b).

e) For preparation of ALP lacking N-linked oligosaccharides, cells were pretreate with tuni- camycin (15 μg/ml) for 4 h, and then labeled with each compound as described above in the presence of tunicamycin.

3. Immunoprecipitation of labeled ALP

a) After being labeled with each compound, the cells were separated from the medium, washed twice with Dulbecco's phosphate buffered saline, and lysed in 0.5 ml of the same buffer containing 1% Triton X-100, 0.5% sodium deoxycholate (DOC) and 0.1% sodium do- decyl sulfate (SDS). The cell lysates were mixed with proteinase inhibitors (antipain, elas- tatinal, leupeptin, chymostatin, pepstatin A, phosphoramidon, and Trasylol, 10 μg each/ ml), sonicated for 2 min, and centrifuged at 15,000 x g for 30 min. The resulting super- natants were used for immunoprecipitation.

b) A 2 μl aliquot of preimmune rabbit serum was added to each cell lysate, and the mixtures were incubated at 4°C for 1 h. The samples were mixed with 100 μl of a 10% (w/v) suspen- sion of Pansorbin which had been pretreated as described above, incubated at 4°C for 1 h with shaking, and centrifuged for 3 min at 12,000 x g in a microcentrifuge. The super- natants were incubated with 3 μl of anti-(placental ALP) serum at 4°C for 4 h, and then with 100 μl of the Pansorbin suspension at 4°C over night.

c) The Pansorbin-immunocomplex was then extensively washed with the following solutions: three times with 10 mM Tris-HCl (pH 7.4) containing 0.7% Triton X-100, 0.7% DOC, 0.2% SDS, 0.7% BSA, 150 mM NaCl, and Trasylol (100 units/ml); twice with 10 mM Tris-HCl (pH 7.4), 0.5% Triton X-100, 0.5% DOC, 0.05% SDS, 150 mM NaCl, Trasylol (100 units/ml); twice with 1 M NaCl in solution A; once with 0.5 M NaCl in solution A; twice with 0.2% SDS in solution A; and once with 10 mM Tris-HCl (pH 7.4), 0.1% SDS, 2 mM EDTA.

d) The immunoprecipitates thus washed were finally suspended in 20-60 μl of 62.5 mM Tris- HCl (pH 6.8) containing 1% SDS, 5 mM EDTA, and 1% 2-mercaptoethanol. The samples were boiled for 4 min, centrifuged at 12,000 x g for 10 min to remove Pansorbin, and stored at -20°C until use. When the samples were used for enzyme digestions, the dis- solving buffer was substituted by other ones as shown below.

4. Enzyme digestions of the labeled ALP

Frozen samples of the labeled ALP were thawed, boiled again for 3 min, and digested at $37^\circ C$ with the following enzymes under the indicated conditions (Takami et al. 1988). If necessary, Nonidet P-40 was added, to protect the enzymes being denatured by SDS initially contained in the samples (at least 7-fold excess amounts of SDS).

a) PI-PLC (10 μg/ml) in 20 mM Tris-HCl buffer (pH 7.5) for 6 h (Kominami et al. 1985).

b) N-Glycanase (10 units/ml) in 0.1 M phosphate buffer (pH 8.6) for 18 h (Plummer et al. 1984).

c) Papain (10 μg/ml) in 50 mM Tris-HCl buffer (pH 7.5) containing 5 mM L-cysteine for 1 h (Takami et al. 1988).

Each digestion was terminated by addition of acetone/0.07 M HCl together with 50 μg of carrier hemoglobin, followed by centrifugation at 12,000 x g for 10 min. The resultant precipitates were washed succesively with acetone/0.07 M HCl and ether, and dried. The proteins were dissolved in 60 μl of 62.5 mM Tris-HCl (pH 6.8), 1% SDS, 5 mM EDTA, 1% 2-mercaptoethanol, boiled for 3 min and stored at $-20^\circ C$ until use.

5. Nitrous acid deamination

The labeled ALP was incubated at $25^\circ C$ for 5 h in 0.5 ml of 0.25 M sodium acetate (pH 3.5) in the presence of freshly prepared 0.2 M $NaNO_2$ (Ferguson and Cross 1984; Low et al.

6. SDS-polyacrylamide gel electrophoresis/fluorography

All the samples thus prepared were analyzed by SDS-polyacrylamide gel (9.0%) electrophoresis/fluorography as described previously (Laemmli 1970; Misumi et al. 1986). Exposure times were usually 3 to 10 days depending upon radioactivies of samples.

COMMENTS

In most of the samples prepared under the conditions described here, the labeled ALP could be identified on fluorographs within 7 days of exposure.

The extensive washing step of immunoprecipitates is most important to obtain clean specific bands on fluorographs. Pretreatments of Pansorbin and of samples with preimmune serum are also strongly recommended.

Since RPMI-1640 medium contains about 20 times myo-inositol of Eagle's MEM, the latter medium should be used for labeling experiments with [^{3}H]inositol. The incorporation of [^{3}H] inositol into ALP, however, was found to be much less than that of other compounds.

In the presence of tunicamycin, ALP was also effectively labeled with [^{3}H]ethanolamine and ^{3}H-fatty acids. However, the incorporation of ^{3}H labeled sugars (mannose, galactose, and glucosamine) was not clearly demonstrated in the presence of tunicamycin. The presence of carbohydrate components in the anchor could be revealed by treatment with N-glycanase of ALP which had been labeled with ^{3}H sugars in the absence of the drug.

REFERENCES

Cross GAM (1987) Cell 48:179-181
Fatemi SH, Tartakoff AM (1986) Cell 46: 653-659
Ferguson MAJ, Cross GAM (1984) J Biol Chem 259:3011-3015
Ferguson MAJ, Low MG, Cross GAM (1985) J Biol Chem 260:14547-14555
Ikezawa H, Yamanegi M, Taguchi R, Miyashita T, Ohyabu T (1976) Biochim Biophys Acta 450:154-164
Ito F, Chou JY (1984) J Biol Chem 259:2526-2530
Kohler PO, Bridson WE (1971) J Clin Endocrinol Metab 32:683-687
Kominami T, Miki A, Ikehara Y (1985) Biochem J 227:183-189
Laemmli UK (1970) Nature 227:680-685
Low MG, Finean JB (1977) Biochem J 167:281-284
Low MG, Ferguson MAJ, Futerman AH, Silman I (1986) TIBS 11:212-215
Low MG, Futerman AH, Ackermann KE, Sherman WR, Silman I (1987) Biochem J 241:615-619
Miki A, Tanaka Y, Ogata S, Ikehara Y (1986a) Eur J Biochem 160:41-48
Miki K, Ogata S, Misumi Y, Ikehara Y (1986b) Biochem J 240:691-698
Misumi Y, Misumi Y, Miki K, Takatsuki A, Tamura G, Ikehara Y (1986) J Biol Chem 261:11398-11403
Plummer TH Jr, Elder JH, Alexander S, Phelan AW, Tarentino AL (1984) J Biol Chem 259:10700-10704
Takami N, Ogata S, Oda K, Misumi Y, Ikehara Y (1988) J Biol Chem 263:3016-3021
Tse AGD, Barclay AN, Watts A, Williams AF (1985) Science 230:1003-1008

Reductive Radiomethylation and Amino Acid Analysis. A Sensitive Procedure to Identify Amine Components in Glycolipid Anchors of Membrane Proteins.

R. Haas and T.L. Rosenberry

Department of Pharmacology, Case Western Reserve University School of Medicine, Cleveland, Ohio 44106, USA

INTRODUCTION

Reductive radiomethylation combined with amino acid analysis is a highly sensitive, flexible, and powerful technique for labeling and identifying amines in proteins. It has proved particularly useful in the study of proteins with glycophospholipid anchors because it also results in the labeling of free amine groups in the anchors and in the identification of ethanolamine and glucosamine as typical anchor components. Reductive radiomethylation of proteins with formaldehyde and $NaCNBH_3$ was first studied systematically by Jentoft and Dearborn (1979). The reaction proceeds readily at pH 7.0 and thus is less likely to result in protein denaturation than earlier procedures with formaldehyde and $NaBH_4$ which required more basic conditions (reviewed by Means 1977). Reductive radiomethylation converts the amines of a protein to their mono- and dimethylated derivatives and, therefore, labels predominantly the lysine residues. At low formaldehyde concentrations, however, the more reactive alpha-amino group of a protein N-terminus is labeled preferentially (Sherman et al. 1983). The methylated amines are stable to standard protein hydrolysis in 6 N HCl, and an amino acid analyzer elution schedule has been developed which can identify the radiomethylated derivatives of all the common amino acids (Haas and Rosenberry 1985). Application of the technique to purified human erythrocyte acetylcholinesterase resulted in the first identification of non-amino acid components in the anchor of this protein (Haas and Rosenberry 1985; Haas et al. 1986; Rosenberry et al. 1986). Because of its sensitivity in identifying the unique glycophospholipid constituents ethanolamine and glucosamine, and in quantitating their amounts, reductive radiomethylation has been extremely useful in the study of a number of other glycophospholipid-anchored proteins including the complement-convertase ininibitor DAF (Medof et al. 1986) and the major murine thymocyte surface antigen Thy-1 (Fatemi et al. 1987). Radiomethylation also complements another exogenous radiolabeling procedure for anchor components involving the photoaffinity label 3-(trifluoromethyl)-3-(*m*-

[^{125}I]-iodophenyl)diazirine ([^{125}I]TID), a reagent specific for anchor lipid (Roberts and Rosenberry 1986; Roberts et al. 1987).

MATERIALS

Equipment

1. Amino acid analyzer: Beckman Model 119CL
   (Other models including HPLC-based systems presumably could easily be adapted)
2. Scintillation counter with two isotope channels
3. Fraction collector with drop counter
4. Protein hydrolysis equipment
5. Speedvac concentrator (Savant)

Reagents

1. Radiomethylation:
   [^{3}H]HCHO, [^{14}C]HCHO, NaCNBH$_3$ (Aldrich, recrystallized)
2. Amino acid analysis:
   Hydrolysis reagents (HCl, mercaptoethanol, argon)
   Analyzer reagents (buffers, ninhydrin)
   Scintillation cocktail

EXPERIMENTAL PROCEDURES

Reductive methylation

Separate 0.05-5.0 ml solutions of protein (typically 0.02-2.4 mg/ml in 20-200 mM sodium phosphate, pH 7.0), appropriate amino acid or amine standards (0.5 mM), and a control free of added amines are incubated with 0.01-100 mM radiolabeled HCHO and 50 mM NaCNBH$_3$ at 37°C for 15 min. Neither high ionic strength nor non-ionic detergents appear to affect the reaction significantly.

Removal of unincorporated radioactivity

Three methods are useful depending on the sample.

a) Dialysis (e.g., overnight with 4 changes of 1 liter of 20 mM sodium phosphate, pH 7.0, at 4^oC). HCHO substantially equilibrates within 1 h. This is the simplest method for proteins and other non-dialyzable species. Small peptide fragments containing glycophospholipid anchors can also be dialyzed safely if they are in a solution containing a non-ionic detergent like Triton X-100 because the anchor binds to the non-dialyzable detergent micelles.

b) Vacuum drying on the Speedvac (for non-dialyzable samples like standard amines). Two dry ice-isopropanol traps in series containing a small amount of water to adsorb volatile radiolabeled reaction products (typically 10^8 cpm) is recommended.

c) Repurification of radiomethylated protein by column or affinity chromatography (generally best as a second step after dialysis).

Hydrolysis

The sample is dried in an acid-cleaned hydrolysis tube. Ultrapure (Baker) 6 N HCl (1 ml) and 6 μl mercaptoethanol are added, and the tube is sealed after four cycles of evacuation and refilling with argon. After hydrolysis at 115^oC for 16 h, acid is removed on the Speedvac. The sample is resuspended in 0.2 M sodium citrate, pH 2.20 (0.5 ml), precipitate is removed by centrifugation, and the sample is injected into the amino acid analyzer.

Amino acid analysis

Conventional amino acid analysis in the Beckman 119CL analyzer utilizes W-3H resin and involves 0.2 M sodium citrate buffers in a 3-step elution: pH 3.23; pH 3.95, 0.2 M NaCl; pH 6.52, 0.8 M NaCl. Times for each step are adjusted to optimize resolution of amino acid standards (Haas and Rosenberry 1985) detected by ninhydrin in methylcellosolve. The sample is injected in 0.2 M sodium citrate (pH 2.20) onto a column (0.6 x 22 cm) equilibrated with pH 3.23 buffer. At the end of the run, the analyzer column is rinsed with 0.2 N NaOH. This protocol will resolve methylated ethanolamine and glucosamine peaks, but most methylated amino acids co-elute in the early 16-24 min region. To resolve them, a 4-buffer elution protocol must be used in which the analyzer column is equilibrated and eluted with the pH 2.20 buffer for 30 min before initiation of the usual 3-buffer cycle. The fourth buffer is conveniently introduced into the analyzer through a 3-way valve inserted in the pH 3.23 buffer line. Non-radioactive amino acids are quantitated in conventional fashion with ninhydrin. To monitor the radioactivity of the column effluent, 0.4 ml (14-drop) to 1 ml (35-drop) fractions of the post-colorimeter effluent are scintillation counted with 5 ml NEF-963 scintillation cocktail (New England Nuclear) per fraction. Each run corresponds to 150-300 fractions.

Interpretation of data

In general, radiomethylation of an amine generates two separate peaks of radioactivity on the analyzer which correspond to the mono- and dimethylated derivatives. Neither derivative reacts significantly with ninhydrin.

1. Peak identification

A radiolabeled analyzer peak can be identified by cochromatography with a standard which has been radiomethylated with a different isotope. A mixture of ^{14}C methylated sample and >10-fold excess of ^{3}H methylated standard is recommended because ^{14}C cpm are obtained unambiguously in the presence of excess ^{3}H cpm in dual channel scintillation counting. Ninhydrin-stained peaks are also convenient reference points. On our system, for example, radiolabeled dimethylethanolamine is close to ninhydrin-stained phenylalanine, and dimethylglucosamine is close to serine. A few radiolabeled artifact peaks are produced as byproducts of the reductive methylation reaction and can be identified in control reactions without added amines.

2. Calculation of methylation stoichiometry

The specific radioactivity of formaldehyde is determined by radiomethylation and analysis of a measured quantity of standard amino acid (Haas and Rosenberry 1985). We typically observe 70 cpm/pmol for [^{14}C]HCHO listed at 40 Ci/mol (ICN Radiochemicals). The total quantity in pmol of a radiolabeled amine is given by the sum of the cpm in the monomethyl peak plus one-half the cpm in the dimethyl peak divided by the specific radioactivity. The quantity of protein in pmol in the hydrolyzate can be estimated simultaneously from the ninhydrin staining of selected amino acids (Ala, Val, Leu, Tyr, Phe, and Arg are among those which remain accurate in the 4-buffer system) if the amino acid composition and the molecular weight of the protein are known (Haas and Rosenberry 1985). Quenching of cpm by ninhydrin color development does not appear to be significant but should be monitored, if possible, on the scintillation counter.

COMMENTS

Reagents

$[^3H]$- and $[^{14}C]$HCHO from a number of suppliers were used without further purification and gave similar results, although some batches showed evidence of partial degradation of HCHO and other batches appeared to contain contaminants which reduced methylation rates. The sensitivity of methylated amine detection (at present about 1 pmol (Haas et al. 1986)) is limited by the specific activity of available radiolabeled HCHO (currently 100 Ci/mol for $[^3H]$HCHO and 40 Ci/mol for $[^{14}C]$HCHO). Since this corresponds to about 0.2% isotopic purity for ^3H, the sensitivity would be improved in theory to 2 fmol with carrier-free $[^3H]$HCHO. NaCNBH$_3$ must be repurified by recrystallization from acetonitrile before use (Jentoft and Dearborn 1979). Solutions made from this very hygroscopic material are stable for only one day.

Labeling reaction

For nearly stoichiometric labeling, the concentration of radiolabeled formaldehyde should be at least 10-fold in excess over that of free amine groups. Under the reaction conditions employed, 10 mM HCHO is usually sufficient to label all amine groups to their fully dimethylated state. A higher concentration (100 mM) is advisable if competing buffer amines or inhibitory reagents such as sodium dodecylsulfate (Jentoft and Dearborn 1979) are present. Lower concentrations (0.2 mM) preferentially label the Nα-amino group of protein N-termini (Haas and Rosenberry 1985). For amino acid and amine standards, 2 mM HCHO gives useful amounts of mono- and dimethylated derivatives. If a lower specific radioactivity is acceptable in large scale radiomethylations, unlabeled HCHO can be added to the reaction mixture to meet the total HCHO concentration requirement.

Interpretation of results

More than 30 distinct peaks can be resolved and identified in a single analyzer run (Haas and Rosenberry 1985). While in general ^3H and ^{14}C methylated amino acid standards cochromatograph, a difference of one 0.4 ml fraction observed reproducibly between the peaks of $[^3H]$- and $[^{14}C]$dimethylethanolamine standards may represent an isotope effect (Haas et al. 1986). Recoveries of methylated amino acids after hydrolysis are not significantly different from those of non-methylated amino acids (Haas and Rosenberry 1985). Under the standard protein hydrolysis conditions, however, both methylated glucosamine and methylated galacto-samine do break down partially to the same peak "X_3" (Haas et al. 1986).

Further applications

Radiomethylated polypeptide mixtures may be resolved by polyacrylamide gel electro-phoresis in sodium dodecylsulfate, and individual labeled bands may be cut from the gel, hydrolyzed and analyzed for radiomethylated amine content (Gnagey et al. 1987). Gel slices ($\leqslant 1$ cm^2) are minced and dried (Speedvac) in the hydrolysis tube before hydrolysis with 3 ml of 6 N HCl and analysis as outlined above. Contaminant amines in the gel slice obscure the ninhydrin trace and prevent accurate protein determination, but radiomethylated amines are recovered uniformly in 50-70% yield.

Radiomethylation can be combined with the manual Edman degradation reaction (Black and Coon 1982) to obtain protein sequence information. Although dimethylation blocks sub-sequent Edman reactions, radiomethylation and analysis following n Edman reaction cycles can be used to identify the $n + 1$ amino acid residue. This procedure is useful for sequencing small peptide fragments attached to glycolipid anchors (e.g., the fragment produced by pa-pain digestion of human erythrocyte acetylcholinesterase (Haas et al. 1986)), because these fragments have solubility properties which are unsuitable for solid phase Edman sequencing. The procedure also is useful for sequencing mixtures of polypeptides since the individual radiomethylated band of interest may be excised and analyzed as outlined above (Haas et al. 1988).

[^{3}H]Ethanolamine has been used as a biosynthetic precursor for radiolabeling the glyco-phospholipid anchors of DAF and Thy-1 in cultured cells (Medof et al. 1986; Fatemi et al. 1987). Hydrolysis and amino acid analysis confirmed that the label was incorporated exclu-sively as [^{3}H]ethanolamine. Methylation of the immunoprecipitated proteins from the labeled cell extracts with *non*-radioactive HCHO prior to hydrolysis and amino acid analysis permit-ted determination of the fraction of ethanolamine incorporated with a free unblocked amino group (Fatemi et al. 1987).

REFERENCES

Black SD, Coon MJ (1982) J Biol Chem 257:5929-5938
Fatemi SH, Haas R, Jentoft N, Rosenberry TL, Tartakoff AM (1987) J Biol Chem 262:4728-4732
Gnagey AL, Forte M, Rosenberry TL (1987) J Biol Chem 262:13290-13298

Haas R, Rosenberry TL (1985) Analyt Biochem 148:154-162

Haas R, Brandt PT, Knight J, Rosenberry TL (1986) Biochemistry 25:3098-3105

Haas R, Marshall TL, Rosenberry TL (1988) Biochemistry (in press)

Jentoft N, Dearborn DG (1979) J Biol Chem 254:4359-4365

Means GE (1977) In: Hirs CHW, Timasheff SN (eds) Methods in Enzymology, Vol 47, Academic Press New York London, pp 469-478

Medof ME, Walter EI, Roberts WL, Haas R, Rosenberry TL (1986) Biochemistry 25:6740-6747

Roberts WL, Rosenberry TL (1986) Biochemistry 25:3091-3098

Roberts WL, Kim BH, Rosenberry TL (1987) Proc Natl Acad Sci USA 84:7817-7821

Rosenberry TL, Roberts WL, Haas R (1986) Fed Proc 45:2970-2975

Sherman G, Rosenberry TL, Sternlicht H (1983) J Biol Chem 258:2148-2156

Analytical and Preparative Phase Separation of Glycolipid-Anchored Membrane Proteins in Triton X-114 Solution

C. Bordier

Biokema SA, 2 ch. de la Chatanerie, CH-1023 Crissier-Lausanne, Switzerland

INTRODUCTION

Non-ionic detergents are widely used for the extraction and during the purification of amphiphilic membrane proteins. Above the critical micelle concentration (CMC) of the detergent, the amphiphilic proteins are found in solution as mixed micelles with the detergent. In contrast, hydrophilic proteins show little or no interactions with non-ionic detergents (Helenius and Simons 1975).

The solubility behaviour of a dilute solution of Triton X-114^{TM} (TX-114) under physiological salt and pH conditions is strongly dependent upon the temperature of the sample. At low temperatures, the detergent forms clear micellar solutions, but above 20°C, a temperature called the cloud point of TX-114, detergent micelles form larger, turbid aggregates and ultimately coalesce to form a separate phase in the sample. The detergent-enriched phase contains approximately 11-13% (w/v) TX-114, and the detergent-depleted phase contains detergent slightly above the CMC (Bordier 1981).

The principle of the phase separation procedure to separate amphiphilic from hydrophilic proteins is based on the concept that above the cloud point, amphiphilic proteins associated with micelles of TX-114 will aggregate with the bulk of the detergent, whereas soluble hydrophilic proteins will remain in the detergent-depleted phase (Bordier 1981).

Glycosyl-phosphatidylinositol (G-PI)-anchored membrane proteins, such as the variant surface glycoproteins of *Trypanosoma brucei*, are amphiphilic membrane proteins with a small and clearly delineated hydrophobic domain (the fatty acids of the lipid anchor) and a large hydrophilic domain (the polypeptide and its associated glycans) (Ferguson et al. 1985). As such, G-PI-anchored membrane proteins may be expected to partition with the detergent during phase separation. In contrast, a G-PI protein from which the hydrophobic domain has been chemically or enzymatically removed (or which, for some reason, failed to receive the anchor during its biosynthesis) would behave as an hydrophilic molecule and partition into the detergent-depleted aqueous phase.

The original phase separation method was described as an analytical tool to distinguish amphiphilic from hydrophilic proteins (Bordier 1981). In this paper, I shall describe a preparative method using extraction and phase separation in TX-114 solution in the purification of a G-PI-anchored protein as well as an analytical method for the assay of G-PI-specific phospholipases. These two applications have been previously published in an abreviated form (Bouvier et al. 1985; Bordier et al. 1986; Etges et al. 1986a,b; Stieger et al. 1986; Ward et al. 1987).

EXPERIMENTAL PROCEDURES

A. PREPARATIVE PHASE SEPARATION OF THE PROMASTIGOTE SURFACE PROTEASE OF *LEISHMANIA MAJOR* (A G-PI-ANCHORED PROTEIN)

Materials

Waterbath at $37^{\circ}C$; ice

Centrifuge, $4^{\circ}C$; tabletop centrifuge, unrefrigerated

40 ml centrifuge tubes

10 mM Tris-HCl, 150 mM NaCl, pH 7.5 (TBS)

Triton X-114 stock solution at 12% w/v in TBS 4 x precondensed (see note below)

Approximately 5×10^{11} washed *Leishmania major* promastigotes

(Optional: DEAE cellulose or Fractogel TSK DEAE-650 (S), chromatography columns, peristaltic pumps, fraction collectors, additional buffers)

Triton X-114 precondensation (Bordier 1981, miniprint section)

Hydrophilic material contained in commercially available TX-114 which will interfere with the phase separation procedure can be eliminated by a few phase separations in the absence of biological material.

Dissolve thoroughly 30 g TX-114 in 1 liter of the desired buffer on ice.

Add a trace of bromophenol blue to stain the detergent phase.

Warm the solution to $37^{\circ}C$ and centrifuge.

Discard the upper phase, and redissolve the detergent-enriched phase in buffer on ice.

Repeat 3-4 times.

Store the detergent-enriched phase as stock at $4^{\circ}C$, or freeze for longer storage.

TX-114 forms a phase containing ~12% detergent when condensed at $37^{\circ}C$ in the presence of 150 mM NaCl.

Methods

1. Resuspend cells in 100 ml TBS and add 20 g Triton X-114 stock
   (final TX-114 concentration = 2%).
2. Incubate on ice for 10 min with occasional stirring.
3. Centrifuge insoluble material at 10,000 x g at $4^{\circ}C$ for 10 min.
4. Decant supernatant and discard cold pellet fraction.
5. Warm supernatant in $37^{\circ}C$ bath. Measure temperature <u>in the tube</u> (>$35^{\circ}C$).
6. Centrifuge turbid solution in the tabletop centrifuge at approximately 1000 x g
   at room temperature for 10 min.
7. Remove upper aqueous phase (detergent depleted). Proceed with lower detergent-
   enriched phase.

7b. (optional)
   Add 100 ml ice-cold TBS to detergent phase and redissolve on ice; repeat steps 3 - 7.

At this stage, the protease is recovered in a solution containing TBS and approximately 12% TX-114. In our hands, the best procedure to eliminate the large excess of detergent is an anion exchange chromatographic step on DEAE (i.e. DEAE cellulose or, better, Fractogel TSK DEAE-650 (S) because of its enhanced flow characteristics with the viscous Triton-enriched sample).

The sample for chromatography is prepared as follows:

8. Measure the volume of the detergent phase and add 12 volumes of 10 mM Tris-HCl,
   pH 8.0 (TB) containing 0.5% Triton X-100. Dissolve on ice and apply to anion exchanger
   (TX-100 inhibits the clouding of TX-114 at room temperature).
9. Wash column with TB containing 0.03% TX-100 or 2.2 mM lauryldodecylamine N-oxide
   (LDAO, a non-ionic detergent which, unlike the Triton detergents, does not absorb at
   280 nm).
10. Elute with a linear gradient of NaCl in the same buffer.

B. AN ASSAY FOR THE GLYCOSYL-PHOSPHATIDYLINOSITOL-SPECIFIC PHOSPHOLIPASE C OF *TRYPANOSOMA BRUCEI* (Ward et al. 1987)

Materials

Waterbaths set at 30° and 37°C

Conical 1.5 ml Eppendorf-type tubes and appropriate centrifuge, room temperature

Iodine-labeled promastigote surface protease of *Leishmania major*, amphiphilic form ($[^{125}I]$-aPSP)

Triton X-114, 4 x precondensed stock at 12% w/v in TBS

10 mM Tris-HCl, 150 mM NaCl, pH 8.0 (TBS) containing 0.05% TX-100

Method

1. Dissolve VSG lipase in TBS in a final volume of 10 μl.
2. In an equal volume of the same buffer, add ~1000 cpm of $[^{125}I]$-aPSP.
3. Incubate at 30°C for 30 min.
4. Stop reaction by adding 600 μl of ice-cold 2% TX-114 in TBS.
5. Mix thoroughly and incubate at 37°C for 2 min to induce clouding.
6. Separate phases by centrifugation at 15,000 g for 1 min.
7. Carefully remove upper phase, avoiding any disturbance of the interface.
8. Measure the radioactivity recovered in the aqueous and in the detergent-enriched phases.
9. Calculate the percentage of digestion by lipase.

This assay is simple but necessitates the availability of radiochemically pure labeled PSP. As an alternative, when the G-PI bound substrate is an enzyme, it is possible to measure its partition after lipase digestion and phase separation by an enzyme assay. This approach was used with acetylcholinesterase of *Torpedo* (Stieger et al. 1986), with the promastigote surface protease of *Leishmania* (Etges et al. 1986a) and with bovine alkaline phosphatase (Ballaman pers. com.).

REFERENCES

Bordier C (1981) J Biol Chem 256:1604-1607

Bordier C, Etges R, Ward J, Turner MJ, Cardoso de Almeida ML (1986) Proc Natl Acad Sci USA 83:5988-5991

Bouvier J, Etges R, Bordier C (1985) J Biol Chem 260:15504-15509
Cardoso de Almeida ML, Turner MJ (1983) Nature 302:349-352
Etges R, Bouvier J, Hoffman R, Bordier C (1985) Mol Biochem Parasitol 14:141-149
Etges R. Bouvier J, Bordier C (1986a) J Biol Chem 261:9098-9101
Etges R, Bouvier J, Bordier C (1986b) EMBO J 5:597-601
Ferguson MAJ, Low MG, Cross GAM (1985) J Biol Chem 260:14547-14555
Helenius A, Simons K (1975) Biochim Biophys Acta 415:29-79
Stieger A, Cardoso de Almeida ML, Blatter MC, Brodbeck U, Bordier C (1986) FEBS Lett 199: 182-186
Ward J, Cardoso de Almeida ML, Turner MJ, Etges R, Bordier C (1987) Mol Biochem Parasitol 23:1-12

Assay and Purification of PI-Specific Phospholipase C from Bacillus cereus Using Commercially Available Phospholipase C

S. Stieger and U. Brodbeck

Institut für Biochemie und Molekularbiologie, Universität Bern, Bühlstrasse 28, CH-3012 Bern, Switzerland

INTRODUCTION

Phosphatidylinositol-specific phospholipase C (PI-PLC) from different sources releases a number of membrane proteins which are anchored to the membrane by a glycosyl-phosphatidylinositol (G-PI) moiety attached to the C-terminus (for review see Low 1987). A typical G-PI-anchored protein is dimeric acetylcholinesterase (AChE) from the electric organ of *Torpedo*, human and bovine erythrocyte membranes, and *Drosophila* heads (Futerman et al. 1983; Taguchi and Ikezawa 1987; Rosenberry et al. 1986; Gnagey et al. 1987). Except mfAChE from human erythrocytes, all these enzymes can be converted to a hydrophilic form (sAChE) by treatment with PI-PLC from *Staphylococcus aureus* (Futerman et al. 1983), *Bacillus thuringiensis* (Taguchi and Ikezawa 1987), *Bacillus cereus* and from *Trypanosoma brucei* (Stieger et al. 1986). Until now, these lipases are not commercially available in pure form.

Here, we describe a simple and rapid assay for PI-PLC using as a substrate mfAChE from readily available sources such as flounder body muscle or bovine erythrocyte membranes. The conversion of mfAChE to sAChE was assessed by phase partitioning using Triton X-114 (Bordier 1981). In this system, amphiphilic proteins are recovered in the detergent phase whereas hydrophilic ones partition into the aqueous phase. Further, we present a method for purification of PI-PLC by affinity chromatography using commercially available PLC from *B. cereus* as starting material. For this purpose, an organomercurial agarose support was used with a high capacity for selectively purifying SH-containing proteins.

Abbreviations. PI-PLC, phosphatidylinositol-specific phospholipase C; AChE, acetylcholinesterase; mfAChE, membrane form AChE; sAChE, soluble AChE; G-PI, glycosyl-phosphatidylinositol

EXPERIMENTAL PROCEDURES

A. ASSAY OF PI-PLC

Materials and reagents

Bovine red cell membrane AChE (Sigma, cat. # C 5021)

Frozen filets of flounder muscle (local food store)

10 mM Tris-HCl, 144 mM NaCl, pH 7.4

Triton X-100

Virtis homogenizer

Sorvall Superspeed centrifuge with 8 x 50 ml rotor

AChE assay solution:

1 mM acetylthiocholine iodide, 0.25 mM 5,5'-dithiobis (2-nitrobenzoic acid), and

0.1 % Triton X-100 in 100 mM phosphate buffer, pH 7.4

Photometer with recorder

Triton X-114, stock solution of 12% (w/v) in 10 mM Tris-HCl, 144 mM NaCl, pH 7.4,

4 x precondensed as described in the chapter of C. Bordier (this volume)

Thermoblock at 37°C (Eppendorf Thermostat 5320)

Eppendorf centrifuge

Methods

mfAChE from the body muscle of flounder or from bovine erythrocyte membranes was used as a substrate for PI-PLC. Conversion of mfAChE to sAChE served as a measure for PI-PLC activity. sAChE and mfAChE were separated by phase partitioning in Triton X-114.

1. Extraction of mfAChE from flounder body muscle

Flounder filets (stored at -20°C) were homogenized with 4 volumes of 10 mM Tris-HCl, 144 mM NaCl, pH 7.4, in a Virtis homogenizer for 2 min. The homogenate was centrifuged at 28,000 x g for 20 min. The supernatant was discarded and the pellet rehomogenized in 10 mM Tris-HCl, 144 mM NaCl, 1% Triton X-100, pH 7.4, and then centrifuged at 28,000 x g for 15 min. The supernatant (detergent extract) contained approximately 3 IU/ml mfAChE (specific activity 1.2 IU/mg protein).

2. Incubation of PI-PLC with AChE

15 μl of detergent extract from flounder body muscle or bovine erythrocyte membrane AChE (approximately 3 IU/ml in 10 mM Tris-HCl, 144 mM NaCl, pH 7.4, specific activity 0.5 IU/mg protein) were incubated at 25°C for 1 h with 5 μl of sample containing PI-PLC in capped Eppendorf tubes (1.5 ml).

3. Phase separation in Triton X-114

0.2 ml of an ice-cold solution of 4% Triton X-114 in 10 mM Tris-HCl, 144 mM NaCl, pH 7.4, were added to the Eppendorf tubes, and AChE activity was measured (= total activity). Phase separation was carried out by incubating at 37°C for 2 min, and subsequently centrifuged in an Eppendorf centrifuge for 1 min. AChE activity was measured in the aqueous supernatant. Conversion of mfAChE to sAChE was calculated as follows:

$$\% \text{ AChE conversion} = \frac{\text{AChE in aq phase} \ \text{x} \ v_{aq}}{\text{AChE total} \ \text{x} \ v_{tot}} \ \text{x} \ 100$$

In these conditions, v_{tot} = 0.220 ml, v_{aq} = 0.152 ml determined by weighing the aqueous phase.

Figure 1 shows the conversion of mfAChE from flounder body muscle and from bovine erythrocytes to the hydrophilic form (sAChE) in dependence of the PLC-concentration.

4. Assay of AChE activity

AChE activity was measured at room temperature by the method of Ellman et al. (1961). 30 μl of sample were mixed in a cuvette with 3 ml of assay solution, and the reaction was followed spectrophotometrically by the increase in absorbance at 412 nm. Enzyme activity (IU/ml) was calculated using a molar absorption coefficient of 13,600 for the nitrobenzoate anion produced in the reaction.

$$\frac{\text{IU}}{\text{ml of enzyme solution}} = \frac{\Delta A_{412}}{\text{minute}} \ \text{x} \ \frac{\text{FV}}{\text{SV}} \ \text{x} \ \frac{7.35}{100}$$

in which FV equals the final volume (3.03 ml), and SV the sample volume (0.03 ml) in the assay.

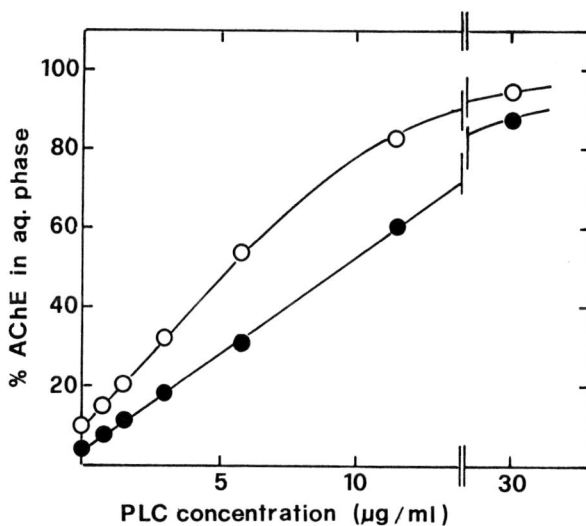

Fig. 1. Conversion of amphiphilic mfAChE to hydrophilic sAChE by PI-PLC assessed by phase separation in Triton X-114. mfAChE from a detergent extract of flounder body muscle (●), and pure mfAChE from bovine erythrocytes (o) were incubated with varying concentrations of crude PLC from *B. cereus*. After 1 h at 25°C, the samples were subjected to phase separation in Triton X-144, and the partition of AChE activity was measured. The sum of activities in both phases remained constant.

B. PURIFICATION OF PI-PLC FROM *B. CEREUS*

Materials and reagents

PLC from *B. cereus* (Sigma, cat. # P 6135 or FLUKA, cat. # 79484)

Affi-Gel 501 (Bio Rad)

Buffer A: 10 mM Tris-HCl, 50 mM NaCl, pH 7.4

Buffer B: 10 mM Tris-HCl, 50 mM NaCl, 10 mM dithiothreitol, pH 7.4

Column (0.8 x 4 cm)

Peristaltic pump

Fraction collector

Aquacide III (Calbiochem)

BCA protein assay kit (Pierce)

Methods

Affi-Gel 501 (2 ml) was packed into a column (0.8 x 4 cm) and equilibrated with buffer A. 1 ml of PLC (1.5 mg protein/ml) in buffer A was applied, and then the column was washed with 100 ml buffer A. After washing, PI-PLC was eluted with 20 ml of dithiothreitol containing buffer B. After application of buffer B, one fraction of 1 ml was collected, and the elution stopped for 30 min. Thereafter, elution was continued slowly (10 ml/h), and fractions of 0.5 ml were collected.

The fractions were assayed for PI-PLC activity as described above. The PI-PLC containing fractions were pooled, filled in a dialysis tube, concentrated in Aquacide III and dialyzed against buffer A.

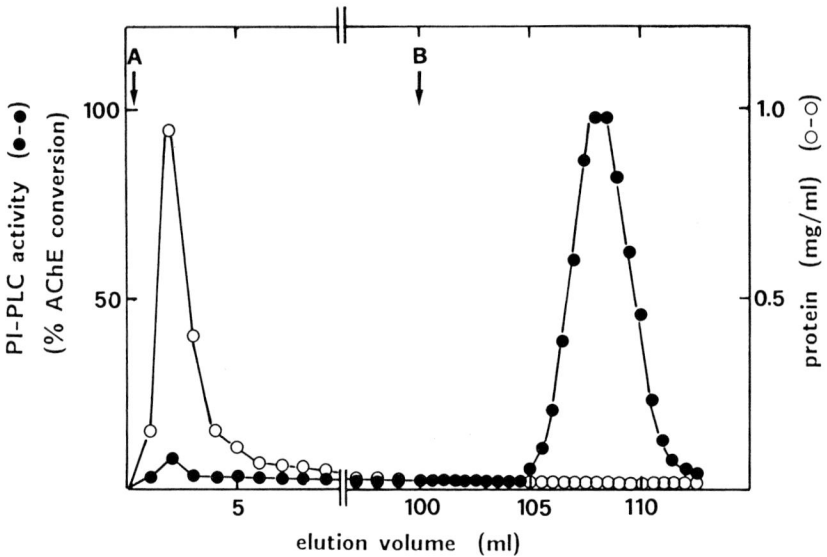

Fig. 2. Affinity purification of PI-PLC from *B. cereus* with organomercurial agarose (Affi-Gel 501). The column was washed with 100 ml of 10 mM Tris-HCl, 50 mM NaCl, pH 7.4 (buffer A). Then PI-PLC was eluted with 10 mM dithiothreitol, 10 mM Tris-HCl, 50 mM NaCl, pH 7.4. PI-PLC activity was measured by the AChE conversion assay using AChE from flounder body muscle, and protein was determined with the BCA protein assay kit.

The yield of PLC activity and the purification factor were determined as follows: Protein was measured with the BCA protein assay kit in the crude and the purified PLC preparation. Then, PI-PLC activity was measured by the AChE conversion assay as a function of protein added to the incubation mixture as shown in Figure 1 for crude PLC. The recovery of PI-PLC activity after affinity chromatography was 30-40%, and a 8-10-fold purification was achieved.

In addition, crude and purified PLC were tested for their ability to hydrolyze phosphatidylcholine (PC) activity. Purified PI-PLC had lost more than 95% of the PC-PLC activity originally present in the PLC preparation.

COMMENTS

For the PI-PLC assay described in this paper, AChE served as a substrate. The conversion of mfAChE to sAChE by PI-PLC can be followed by measuring enzyme activity after phase separation in Triton X-114. Since AChE can be detected by activity measurement using the highly sensitive Ellman assay, only small amounts of enzyme (approximately 40 mIU or 0.1 pmoles AChE per assay) are required, and it is not necessary to purify AChE. A detergent extract from flounder body muscle was found to be the most suitable source for mfAChE since more than 95% of this amphiphilic AChE partition into the detergent phase, even in the crude extract, while a near quantitative recovery in the aqueous phase was obtained after conversion to the hydrophilic form. An alternate commercially available substrate is crude mfAChE from bovine erythrocyte membranes (Sigma). Although the membrane form of this enzyme partitions to about 80% into the detergent phase, it is still suitable for the PI-PLC assay.

In contrast to other procedures used for purification of PI-PLC from *B. cereus* (Sundler et al. 1978), the affinity chromatography with an organomercurial ligand described in this paper is a very simple and rapid one-step procedure, clearly separating the anchor-splitting activity from other phospholipid cleaving activity.

REFERENCES

Bordier C (1981) J Biol Chem 256:1604-1607
Ellman GL, Courtney DK, Andres V, and Featherstone RM (1961) Biochem Pharmacol 7:88-95
Futerman AH, Low M, Silman I (1983) Neurosci Lett 40:85-89
Gnagey AL, Forte M, Rosenberry TL (1987) J Biol Chem 262:13290-13298
Low MG (1987) Biochem J 244:1-13
Rosenberry TL, Roberts WL, Haas R (1986) Fed Proc 45:2970-2975
Stieger A, Cardoso de Almeida ML, Blatter MC, Brodbeck U, Bordier C (1986) FEBS Lett 199: 182-186
Sundler R. Alberts AW, Vagelos PR (1978) J Biol Chem 253:4175-4179
Taguchi R, Ikezawa H (1987) J Biochem 102:803-811

Use of Phenylsepharose to Discriminate Between Hydrophilic and Hydrophobic Forms of Decay Accelerating Factor

M.A. Davitz

Departments of Environmental Medicine and Pathology, New York University School of Medicine, 550 First Avenue, New York, NY 10016, USA

INTRODUCTION

Decay accelerating factor (DAF) is a membrane protein which binds to complement fragments C3b or C4b (Kinoshita et al. 1986) inhibiting amplification of the complement cascade on cell surfaces (Hoffman 1969; Nicholson-Weller et al. 1982; Pangburn et al. 1983). Recently, DAF has been shown to be anchored to the cell membrane by phosphatidylinositol (PI) Davitz et al. 1986; Medof et al. 1986). Native purified membrane DAF possesses the unusual property of reincorporating into cell membranes (Medof et al. 1984); additionally, it binds tightly to phenylsepharose (Medof et al. 1986). Both reincorporation and binding are probably mediated by the diacylglycerol anchor since DAF, treated with PI-specific phospholipase C (PI-PLC), does not bind to the beads nor does it reincorporate into lipid bilayers (Davitz et al. 1986; Medof et al. 1986). While studying the properties of a glycan-specific phospholipase D in human serum, I developed a novel assay involving binding of the PI-anchored protein to phenylsepharose which discriminated between hydrophobic and hydrophilic forms of DAF (Davitz et al. 1987b). The assay is rapid and, potentially, could be used to screen large numbers of samples.

EXPERIMENTAL PROCEDURES

1. Solutions and materials

a) Phenylsepharose (Pharmacia):
   Prior to use, the beads are washed in PBS containing 0.001% Nonidet P-40 (NP-40).
b) DAF is purified from human erythrocytes by immunoaffinity chromatography (Davitz et al. 1987a).

c) The source of the phospholipase D in this case, is human serum. This enzyme specifically cleaves DAF removing phosphatidic acid from the COOH terminal glycolipid (Davitz et al. 1987b; Low and Prasad 1988).

d) Buffer 1:

50 mM Tris-HCl, pH 7.4, 10 mM NaCl, 2.6 mM $CaCl_2$

e) Phosphate buffered saline (PBS) (Dulbecco)

2. Two-site immunoradiometric assay for DAF

This is performed with monoclonal antibodies as previously described (Kinoshita et al. 1987). The amount of DAF in the various fractions is calculated from a standard curve obtained with purified DAF. This curve in which the specific counts of the revealing antibody is plotted as a function of DAF concentration, is linear up to 250 ng of DAF/ml.

3. Sample data

Total amount of DAF/assay = 218 ng

Assay Condition	Quantity of DAF Converted*	%Conversion
	187ng	86
+serum	179ng	82
	170ng	78
	9.7ng	5.1
-serum	10.7ng	4.9
	10.8ng	5.0

* determined by two-site immunoradiometric assay

4. Protocol

Purified membrane DAF (0.22 μg) is incubated at 37°C for 1 h with 0.1 μl of human serum in buffer 1 (total volume 100 μl). After incubation, 900 μl of PBS is added to the reaction mixture followed by adding 75 μl of phenylsepharose beads. The mixture is incubated at room temperature (20°-24°C) for 1 h, the beads removed by centrifugation in a microcentrifuge at 10,000 rpm for 1 min, and the supernatants assayed for DAF by means of a two-site immunoradiometric assay. The percentage conversion of membrane DAF is calculated as 100 x μg of DAF in the supernatant/0.22 μg.

COMMENTS

As long as a sufficently sensitive assay for the PI-anchored protein exists, it should be possible to use this assay to discriminate between hydrophilic and hydrophobic forms of the molecule. Moreover, if the binding capacity of the phenylsepharose beads exceeds the total amount of protein in the mixture (phenylsepharose binds approximately 15 mg of bovine serum albumin/ml of gel), it should not be necessary to use purified material. Since binding to phenylsepharose is quite rapid, this method could potentially be utilized for screening a large number of samples.

Several points must be kept in mind. Binding to phenylsepharose beads is dependent on both the temperature and salt concentration. I achieved optimal results at room temperature (20°-24°C) in 0.15 M NaCl, pH 7.4. These particular conditions may need to be adjusted for each protein.

Acknowledgment. M.A. Davitz is supported by a Physician Scientist Award (1-NIEHS ES00136)

REFERENCES

Davitz MA, Low MG, Nussenzweig V (1986) J Exp Med 163:1150-1161
Davitz MA, Schlesinger D, Nussenzweig V (1987a) J Immunol Meth 97:71-76
Davitz MA, Hereld D, Shak S, Krakow J, Englund PT, Nussenzweig V (1987b) Science 238: 81-84
Hoffman EM (1969) Immunochemistry 6:391-403
Kinoshita T, Medof ME, Nussenzweig V (1986) J Immunol 136:3390-3395
Kinoshita T, Medof ME, Silber R, Nussenzweig V (1987) J Exp Med 162:75-92
Low MG, Prasad ARS (1988) Proc Natl Acad Sci USA 85:980-984
Medof ME, Kinoshita T, Nussenzweig V (1984) J Exp Med 160:1558-1578
Medof ME, Walter EI, Roberts WL, Haas R, Rosenberry TL (1986) Biochemistry 25:6740-6747
Nicholson-Weller AJ, Burge J, Fearon DT, Weller PF, Austen KF (1982) J Immunol 129: 184-189
Pangburn MK, Schreiber RD, Muller-Eberhard HJ (1983) Proc Natl Acad Sci USA 80:5430-5434

PI-PLC Assay on Nitrocellulose Filter Immobilized Proteins

C. Deregnaucourt and Y. Capdeville

Centre de Génétique Moléculaire, CNRS, F-91190 Gif sur Yvette, France

INTRODUCTION

The major evidence for covalent linkage of a membrane protein to a phosphatidylinositol lipid (Low et al. 1986; Cross 1987) is generally provided by cleavage of the lipid by a phosphatidylinositol-specific phospholipase C (PI-PLC). Generally, this cleavage leads to the unmasking of a particular epitope, the "cross-reacting determinant" (CRD) which includes ethanolamine, sugars, inositol, and phosphate. The CRD was first described (Barbet and Mac Guire 1978) in *Trypanosoma brucei* variant surface glycoproteins and demonstrated to be part of a complex post-translational structure linked, at one side, to the COOH terminus of the protein, and at the other side, to the membrane anchoring lipid. Removal of the lipid is necessary for the antibodies to have access to the CRD (Cardoso de Almeida and Turner 1983).

Phospholipase C assays are generally performed in solution. Here, we report experiments which show the ability of PI-PLC from *Bacillus cereus* to cleave diacylglycerol from denatured and reduced proteins after they have been transferred to nitrocellulose.

These experiments were carried out on a protozoan, *Paramecium primaurelia* which displays antigenic variation. *Paramecium* surface antigens (SAgs, glycoproteins of MW 250-300 kDa) can be isolated under two distinct forms (Capdeville et al. 1987): a membrane-bound form (mSAg) which is myristylated, and a soluble form (sSAg) lacking the lipid and displaying the CRD. The mSAg can be released as sSAg by action of endogenous phospholipase C-like hydrolase and exogenous PI-PLCs from *T. brucei* or *B. cereus*.

Western blot analysis of a partially purified preparation of the soluble form of SAg also shows several polypeptides in the 40-60 kDa range which are recognized by anti-CRD antibodies along with the sSAg. Those glycopeptides are not detected by anti-CRD antibodies in preparations containing the membrane form of the SAg, except after this type of preparation has been incubated with PI-PLC either from *T. brucei* or from *B. cereus*. Thus, the origin of these glycopeptides raised the following questions: are they the result of proteolysis of the SAg occuring after cleavage of the lipid moiety, or do they correspond to distinct proteins with glycosylinositol phospholipid anchors?

The second possibility inferred that such proteins, in their "membrane" form, would be sensitive to the action of PI-PLC, this leading to the unmasking of the CRD. To check this hypothesis, we first separated the proteins present in mSAg preparation by SDS-PAGE. The proteins were then transferred onto nitrocellulose and the filters were incubated with or without PI-PLC, respectively, from *B. cereus* before incubation with anti-CRD antibodies. Results clearly showed that only on the filter incubated with PI-PLC were the glycopeptides detected by anti-CRD antibodies. This contributed to the demonstration that these glycopeptides, now called "cross-reacting glycoproteins" (CRGs), are membrane proteins distinct from the SAg.

EXPERIMENTAL PROCEDURES

Immunoblotting was carried out according to Towbin et al. (1979) with slight modifications (Capdeville et al. 1985). After saturation with bovine serum albumin, the nitrocellulose sheets were incubated at 37°C for 2 h in 10 mM Tris, 150 mM NaCl, pH 7.3, in the presence of 2% (w/v) Triton X-100 and protease inhibitors (0.25 mM phenylmethylsulfonyl fluoride, 0.01 mM leupeptin, 1 mM EDTA), and Phospholipase C (type III) from *B. cereus* (Sigma) at a final concentration of 2 units/ml. The blots were then washed three times for 10 min with phosphate buffered saline, pH 7.5, before incubation with antibodies. Anti-CRD purified antibodies (3 mg/ml) were diluted about 500-fold. These purified antibodies directed against the CRD of *Trypanosoma brucei* were a gift from M.L. Cardoso de Almeida. Peroxydase-conjugated sheep anti-rabbit IgGs (Institut Pasteur, France) were used at the dilution 1:200 with 4-chloro-1-naphtol (Sigma) as chromogenic substance.

COMMENTS

We have shown that PI-PLC from *B. cereus* can act on denatured and reduced molecules immobilized on nitrocellulose. This simple procedure allows detection of the CRD after action of PI-PLC on proteins first isolated and fixed in their lipid-linked form. The main interest of such a procedure is to avoid any proteolysis of the studied proteins which could occur during or after cleavage of the lipid moiety. Indeed, such possible proteolysis must be considered when PI-PLC assay is performed on a crude or partially purified preparation of the "substrate" proteins. Therefore, this procedure gives a direct inventory of membrane proteins anchored through a phosphatidylinositol, provided that these proteins display the CRD after cleavage of the lipid moiety by PI-PLC.

REFERENCES

Barbet AF, MacGuire TC (1978) Proc Natl Acad Sci USA 75:1989-1993

Capdeville Y, Deregnaucourt C, Keller AM (1985) Exp Cell Res 161:495-508

Capdeville Y, Cardoso de Almeida ML, Deregnaucourt C (1987) Biochem Biophys Res Commun 147:1219-1225

Cardoso de Almeida ML, Turner MJ (1983) Nature 302:349-352

Cross GAM (1987) Cell 48:179-181

Low MG, Ferguson MAJ, Futerman AH, Silman I (1986) TIBS 11:212-215

Towbin H, Staehelin T, Gordon J (1979) Proc Natl Acad Sci USA 76:4350-4354

Deacylation and Reacylation of a Pure Enzyme to Demonstrate Covalent Glycerolipid

P.R.V. Nayudu[*], B. Lambrecht[+], and M.F.G. Schmidt[#]

[*]Department of Zoology, Monash University Melbourne, Clayton, Victoria 3168, Australia
[+]Institut fur Virologie, Justus-Liebig-Universität, Frankfurter Strasse 107,
D-6300 Giessen, Federal Republic of Germany
[#]Department of Biochemistry, Faculty of Medicine, Kuwait University, P.O. Box 24923,
13110 Safat, Kuwait, Arabian Gulf

INTRODUCTION

Alkaline phosphatase is a membrane-bound enzyme of several mammalian tissues, and a covalently attached phospholipid has been proposed as its anchor to the membrane bilayer (Low et al. 1986). Attempts to label the enzyme of the mouse intestinal epithelial cell's microvillous membrane in vivo metabolically with [^{3}H]fatty acid were hampered by a small quantity of the enzyme per animal (Low et al. 1986), and by the problem of conversion of the labeled lipid to other metabolites during the long period (about 36 h) required for the molecular differentiation of the active enzyme as the cells migrate from the crypts to the tips of the intestinal villus (Nayudu 1984).

Since this enzyme has been purified to homogeneity in our laboratory (Nayudu 1984), in vitro labeling with [^{3}H]fatty acid was attempted to demonstrate glycerolipid as a covalent structural component of the pure enzyme. This approach may be applicable to any enzyme which has covalent glycerolipid, provided it can be obtained in a pure form in sufficient quantity.

MATERIALS

1. Deacylation

Pure enzyme (~1 mg) in 20 mM HEPES buffer, pH 7.4, containing 0.14 M KCl and 0.5 mM $MgCl_2$
Insoluble beaded agarose-phospholipase A_2 (Sigma, USA)

Octyl-β-D-glucoside 20% solution
Shaker, set-up on a Vortex mixer
Eppendorf centrifuge and screw-cap tubes

2. Reacylation

5 plates (15 cm diameter) of monolayer culture of BHK cells
Buffer, 20 mM HEPES, 0.14 M KCl, 0.5 mM $MgCl_2$, pH 7.4
Dounce homogenizer, tight-fitting
Superspeed centrifuge, SW rotor for 10 ml tubes
[9,10-^{3}H]palmitic acid (New England Nuclear, Boston, USA)

3. Polyacrylamide gel electrophoresis

7% slab gels (20 x 16 x 0.3 cm) with 0.1% SDS in gel buffer
Electrode buffer (25 mM Tris, 0.2 M glycine with 0.1% SDS)
Denaturing sample solubilizer (1.25 ml glycerol, 1.5 ml 2-mercaptoethanol, 4.5 ml 20% SDS, 1.25 ml 1.25 M Tris, pH 6.8, and 1 μl bromophenol blue)
Coomassie brilliant blue stain
Gel dryer

4. Detection of radioactivity

Berthold scanner (LB 2723, LB 2842; Wildbad, FRG) coupled to a computer for data processing
Fluorography of dried gels, Kodak X-R5 film, and photographic facilities

EXPERIMENTAL PROCEDURES

1. Deacylation of pure enzyme (Lambrecht 1983)

To an Eppendorf tube, the following are added:
a) 0.2 ml of pure enzyme (1 mg) in HEPES buffer,

b) 28 μl octyl-β-D-glucoside, and

c) 10 units of insoluble phospholipase A_2.

The tube is capped and shaken at room temperature for 3 h. Subsequently, the insoluble phospholipase is removed by centrifugation in the Eppendorf centrifuge for 15 min. The supernatant is used for the reacylation.

2. Microsomal membranes from BHK cells (Berger and Schmidt 1984)

Cells are scraped from the plates and washed twice in cold HEPES buffer and pelleted in a refrigerated centrifuge at 1000 x g. Packed cells are resuspended in 10 volumes of buffer and homogenized at 4°C in the Dounce homogenizer (40 strokes). The homogenate is centrifuged using an angle rotor at 11,000 x g for 20 min, and the nuclear pellet is discarded. The supernatant is then centrifuged using SW rotor at 150,000 x g for 1 h. The pellets of microsomal membranes are resuspended in 10 volumes of buffer using a loose-fitting homogenizer (5 strokes) to obtain a membrane suspension of about 5 mg protein/ml.

3. Reacylation of deacylated enzyme

To the 0.2 ml supernatant of experiment 1 above, the following additions are made:

a) 0.2 ml of BHK microsomal membranes from experiment 2 above,

b) neutralized solution of ATP to obtain a final concentration of 1 mM, 100 μg phenylmethylsulfonylfluoride, 20% octyl-β-D-glucoside to 1%.

c) 100 μCi [9,10-^{3}H]palmitate is dried in an Eppendorf tube to remove the alcohol, and the reacylating mixture above is carefully transferred to the tube containing the palmitate. The screw-cap is replaced tightly and the tube wrapped in parafilm.

d) The tube is shaken mildly on a vortex shaker for 4 h. The tube is then centrifuged in an Eppendorf centrifuge at 14,000 rpm for 30 min, and the pelleted microsomal membranes separated from the clear supernatant.

4. Control experiment

1 mg pure enzyme not subjected to deacylation with phospholipase A_2 is set up separately exactly as set out in experiment 3. This control is necessary to show that reacylation with radioactive palmitic acid can occur only if the existing fatty acid in the pure enzyme is removed by insoluble phospholipase A_2.

5. Extraction with CMA

(chloroform - methanol - 1 M HCl (2:1:0.025, v/v))

The supernatants experimental (3) and control (4) are processed in an identical manner. 20 volumes of CMA are added to the supernatant in a 15 ml centrifuge tube and mixed carefully. More methanol (1/4 volume) is added to make it into a single phase. The tubes are centrifuged at 5000 rpm, and the precipitated protein carefully removed and redissolved in 0.5 ml buffer. The organic solvent containing most of the radioactivity is disposed of appropriately. The redissolved protein is re-extracted with CMA, and the precipitated protein taken in 10-50 μl of buffer.

6. Polyacrylamide gel electrophoresis (Laemmli 1970)

5-30 μl of redissolved protein sample from experiment 5 is transferred to an Eppendorf tube, the aliquot containing approximately 200-300 μg protein. The samples for electrophoresis should consist of:

a) control, pure untreated enzyme (about 200 μg protein),

b) control, enzyme from experiments 4 and 5, and

c) reacylated enzyme from experiments 3 and 5.

To each, tube 1/2 volume of three times concentrated sample solubilizer is added, and the tubes are placed in a boiling waterbath for 5 min. After cooling, each sample is applied to the gel in 3 slots with protein concentration varying 1:2:4 between slots. Electrophoresis is carried out at constant voltage (50-150 volts) over a duration required to move the bromophenol blue to the bottom of the gel. The gel is then removed and stained with Coomassie brilliant blue, destained, and then dried on the gel dryer.

7. Determination of tritium labeling of the enzyme protein

This is done by scanning the tracks (about 1 cm wide) for each sample on the dry gel using Berthold thin-layer chromatogram scanner coupled to an Apple computer for data processing (Schmidt 1984). Prior to drying, the gel may be treated with Amplify (Amersham, U.K.), in which case the gel may also be fluorographed (Bonner and Laskey 1974) using Kodak X-R5 film.

COMMENTS

The untreated pure enzyme is used as a control sample in order to identify the protein bands constituting the pure enzyme on the electrophoretic gel. The other control of the pure enzyme set-up for reacylation without prior deacylation with insoluble phospholipase A_2 should not show any radioactivity in the enzyme-protein bands because the premise of the experiment is that the phospholipid of the enzyme has intact fatty acids. Evidence of a positive result would, therefore, comprise of radioactive labeling of the enzyme-protein which was first deacylated and subsequently set up for reacylation with the help of acyltransferase activity of the BHK microsomes (Berger and Schmidt 1986).

The validity of the experiment is based on two facts:

a) The extraction of the protein sample with CMA after reacylation removes all lipid except that covalently bound to the protein, and

b) in protein electrophoresis in the presence of SDS when the samples were applied after boiling with the REDUCING SOLUBILIZER; any lipid not covalently bound to the proteins, moves with bromophenol blue.

Both these expectations have been shown to be valid experimentally.

Further confirmation of the covalent linkage of labeled fatty acid may be obtained by alkaline hydrolysis (3 N NaOH, 100°C) of the protein band recovered from the gel and demonstrating the recovery of the fatty acid in the chloroform phase following chloroform-methanol extraction of the hydrolyzed sample.

REFERENCES

Berger M, Schmidt MFG (1984) J Biol Chem 259:7245-7252
Berger M, Schmidt MFG (1986) J Biol Chem 261:14912-14918
Bonner WM, Laskey RA (1974) Eur J Biochem 46:83-88
Laemmli UK (1970) Nature 227:680-685
Lambrecht B (1983) Diploma Thesis, Justus-Liebig-Universität, Giessen, FRG
Low MG, Ferguson MAJ, Futerman AH, Silman I (1986) TIBS 11:212-215
Nayudu PRV (1984) Biochem Internat 8:193-202
Schmidt MFG (1984) EMBO J 3:2295-2300

Incorporation of Rat Brain Thy-1 Antigen into Mouse T Lymphocytes

M. Puklavec and A.F. Williams

MRC Cellular Immunology Unit, Sir William Dunn School of Pathology, University of Oxford, Oxford OX1 3RE, U.K.

INTRODUCTION

Molecules with glycosyl-phosphatidylinositol (G-PI) anchors might be expected to spontaneously incorporate into a cell surface when cultured with viable cells. Thy-1 is one prototype for G-PI anchored molecules and exists as a monomer in association with detergent micelles or as an oligomer containing about sixteen molecules when detergent is removed (Kuchel et al. 1978; Campbell et al. 1981; Tse et al. 1985). In the mouse, T lymphocytes can be triggered into cell division when cell surface Thy-1 is cross-linked with a monoclonal antibody and anti-immunoglobulin second antibodies, and the cells are cultured with phorbol myristate acetate (PMA) (Kroczec et al. 1986). With this in mind, we attempted to incorporate pure rat brain Thy-1 into the cell surface of mouse T lymphocytes with a view to testing triggering via the rat Thy-1. Mouse Thy-1 comes in two allotypic forms named Thy-1.1 and Thy-1.2 which differ by expressing Arg (Thy-1.1) or Gln (Thy-1.2) at residue 90 (Williams and Gagnon 1982). Rat Thy-1 has Arg at residue 89 (which is equivalent to mouse residue 90), and anti-Thy-1.1 antibodies cross-react between rat Thy-1 and mouse Thy-1.1 but do not bind to mouse Thy-1.2 (Mason and Williams 1980). Thus, if rat brain Thy-1 were incorporated into a mouse lymphocyte of Thy-1.2, type triggering could be attemped via the endogenous Thy-1 using an anti-Thy-1.2 monoclonal antibody (MAb), or by the inserted Thy-1 with an anti-Thy-1.1 MAb. In the present studies, the anti-Thy-1.2 MAb used was 30H12 (Ledbetter and Herzenberg 1979), and the anti-Thy-1.1 MAb was MRC OX-7 (Mason and Williams 1980).

Incorporation into erythrocytes of the G-PI anchored form of decay accelerating factor has been reported by simple incubation of cells with the pure molecule but the amounts incorporated were low (Medof et al. 1984). In the present study, incorporation into lymphocytes was facilitated by incubation with the membrane destabilizing agents Na-cholate or polyethylene glycol (PEG).

EXPERIMENTAL PROCEDURES

Preparation of rat thymocytes and mouse T lymphocytes

PVG rat thymocytes were teased into Dulbecco's A+B (DAB) buffer plus 0.2% w/v bovine serum albumin (BSA) with watchmakers forceps, and tissue debris was removed by filtration through cotton wool. Cells were counted on a haemocytometer or by use of a Coulter Counter (Hunt 1987). Balb/c mouse spleen cells were prepared as above and B cells were removed by rosetting with sheep erythrocytes coated with purified rabbit anti-mouse IgG antibody (RAM) (Mason et al. 1987).

Cell culture

All incubations at room temperature or at $37^{\circ}C$ were in RPMI-1640 media plus 10% foetal calf serum (FCS) with 5% CO_2. Mitogenesis experiments were carried out in some cases with PMA at 1 μg/ml and others at 10 ng/ml; the cells showed equal responses with PMA over this range. In some experiments, cells at 1.25×10^6/ml were incubated with anti-Thy-1 antibodies and then washed before setting up in culture with RAM (a fraction which cross-reacts between mouse and rat IgG) at 20 μg/ml and PMA (Kroczec et al. 1986) at 10 ng/ml or 1 μg/ml. In other experiments, anti-Thy-1 antibodies, at 10 μg/ml, were present throughout the culture. In all cases, 2×10^5 cells per well were set up in culture.

Rat brain Thy-1

This was purified with an MRC OX-7 MAb affinity column as in Campbell et al. (1981). Thy-1 was eluted from the column in buffer containing 0.5% deoxycholate, 10 mM Tris-HCl, 0.05 M diethylamine-HCl, pH 11.5, and the eluate was neutralized with glycine. The material was exhaustively dialyzed with 0.1 M NH_4HCO_3, and then with PBS for 24 h prior to use in incorporation experiments, and the molecule can thus be expected to have been in an oligomeric form (Kuchel et al. 1978).

Quantitation of incorporated rat brain Thy-1

This was done using ^{125}I labeled F(ab')$_2$ fragments of MRC OX-7 antibody. Purification and labeling was exactly as in Mason and Williams (1980). Cells into which Thy-1 had been incorporated, were counted and dispensed at 10^6 cells per well in 0.2 ml microtitre wells. [^{125}I]MRC OX-7 F(ab')$_2$ at 20 μg/ml and 1.0 x 10^7 cpm ^{125}I/ml was added in 50 μl DAB/ 0.5% BSA. Cells were incubated at 4^0C for 1 h, and washed four times by centrifugation in DAB/0.2% BSA. The cpm bound was determined and converted into molecules of F(ab')$_2$ MRC OX-7 bound per cell. The data was not corrected for possible cell loss during washing.

RESULTS

Determination of non-lethal levels of membrane destabilizing agents

Sodium cholate and polyethylene glycol (PEG) were investigated. Cells were incubated in DAB/ 0.2% BSA at room temperature for 30 min with various levels of each material, and then washed and assayed for mitogenesis via anti-Thy-1.2 antibody and PMA with the results shown in Table 1. PEG was non-lethal at levels up to 20% (higher levels were not tested) whilst cholate was non-toxic up to 0.25% but completely inhibited mitogenic activity at 0.5% (in another experiment, inhibition of response also occurred at 0.25%).

Table 1. Effect of Na-cholate and PEG on Thy-1 mediated cell division

Percentage agent in pre-treatment		cpm ($\times 10^{-3}$) ^{3}H Thymidine Incorporation	
PEG	Na-Cholate	PEG series	Na-Cholate series
0	0	139	125
1.25	0.03	120	153
2.5	0.06	127	153
5.0	0.125	133	162
10.0	0.25	157	128
20.0	0.5	145	4
Controls			
+PMA-antibody		3	
-PMA+antibody		0.3	

Mouse T lymphocytes were cultured in PEG or Na-cholate in DAB/0.2% BSA at the concentration shown at room temperature for 30 min. The cells were then washed in DAB/0.2% BSA and set up in culture with PMA, anti-Thy-1.2 MAb, and RAM as described in Experimental Procedures.

Incorporation of Thy-1

Unfractioned mouse splenocytes were incubated with rat Thy-1 only, or with Thy-1 plus various levels of cholate and PEG with the results shown in Table 2. Incorporation was assessed by labeling with [^{125}I]MRC OX-7 antibody, and rat thymocytes were labeled as a positive control. The rat cells bound the antibody at levels of about 400,000 molecules per cell (Table 1) which is lower than the expected level of 700,000 molecules per cell (Mason and Williams 1980). The deficit could be due to loss of cells in washing since cells were not counted to correct for this source of error.

Table 2. Integration of rat Thy-1 into Balb/c splenocytes

Treatment	Molecules (x10^{-3}) of ^{125}I MRC OX-7 bound per cell	
	Experiment Number	
	1	2
Buffer (background)	4.9	6.4
Buffer + Thy-1	41	28
+ PEG (%)		
22.5	587	726
18	587	264
9	312	125
4.5	129	
2.2		69
1.8	88	
1.1		50
+ Cholate (%)		
0.125		69
0.05	88	20
0.03		20
Rat thymocytes (no treatment)	312	454

10^7 cells were incubated in 0.1 ml DAB/0.2% BAS at room temperature for 30 min containing the additions shown. The Thy-1 was at 200 μg/ml, and the suspensions were shaken. The cells were washed and set up for binding assays with [^{125}I]F(ab')$_2$ MRC OX-7 at 10^6 cells per well as described in Experimental Procedures.

After incubation with Thy-1 alone at 200 μg/ml, mouse lymphocytes bound [^{125}I]MRC OX-7 at about 30,000 molecules per cell. With PEG, the level of incorporation was greatly increased while Na-cholate additions gave only modest effects. The concentration of cells and Thy-1 was important since with incubation at a 10-fold dilution of cells and Thy-1 (but not PEG), incorporation of Thy-1 was reduced about 6-fold. For routine studies, Thy-1 was incorporated into T cells by incubation with PEG at 10%.

To assess the distribution of incorporated Thy-1 amongst the total cell population, cells were labeled with MRC OX-7 MAb followed by fluorescein-labeled RAM and analyzed with the fluorescence-activated cell sorter. As shown in Figure 1, the mouse T cells were uniformly labeled at a lower level than seen for the rat thymocytes which were labeled as a positive control.

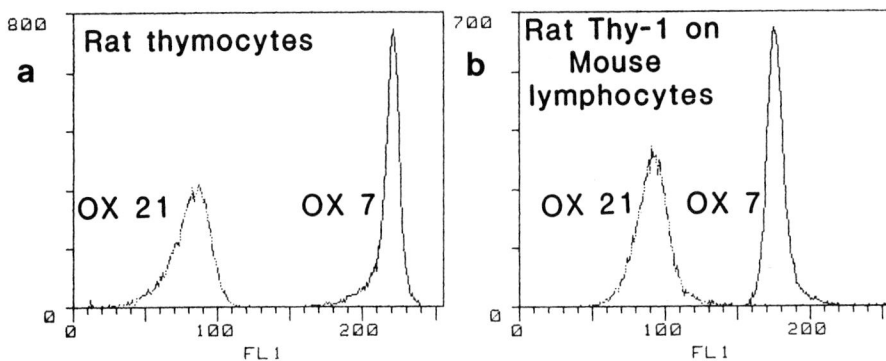

Fig. 1. Labeling of rat Thy-1 on rat thymocytes (**a**) and mouse T lymphocytes with rat Thy-1 incorporated (**b**). Rat brain Thy-1 was incorporated into Balb/c mouse lymphocytes with 10% PEG as described in Table 2. The cells were then washed and incubated with unlabeled MRC OX-7 or MRC OX-21 (control) MAbs in the form of tissue culture supernatant at $4^{o}C$ for 60 min. After washing, incubation with fluorescein RAM (specific for mouse IgG) was at $4^{o}C$ for 60 min, and cells were then analyzed on a Becton Dickinson FACS II flow cytometer. For control Balb/c mouse lymphocytes labeled with MRC OX-7, the profile was as for the MRC OX-21 control.

Stability of Thy-1 incorporation

Rat brain Thy-1 was incorporated into mouse T lymphocytes, and then the cells were washed and cultured at $37^{o}C$ for various time intervals. This was done without any stimulating agents or with stimulation via endogenous Thy-1.2. After culture, the cells were taken and incubated with $[^{125}I]$MRC OX-7 to measure the level of remaining rat Thy-1. As shown in Table 3, Thy-1 was retained in the membrane of both resting and activated cells with little loss. In these experiments, cells were aliquoted into microculture trays for the $37^{o}C$ incubation and then assayed with $[^{125}I]$MRC OX-7 in the same wells without allowance for cell death. Any loss of cells would reduce the apparent recovery of Thy-1.

Table 3. Stability of integration of rat Thy-1 in Balb/c splenocytes

Culture time (hr)	Molecules ($x10^{-3}$) of ^{125}I F(ab')$_2$ MRC OX-7 bound/cell			
	Balb/c splenocytes		Rat thymocytes	
	1	2	1	2
0	147	69	429	
1	98		443	
2	104		443	
3	96		439	
4	86		412	
6	93		429	
8	87		409	
12		82		359
19			370	
36		89		384
18 (4°C)		119	403	384
Buffer	7.5			

Thy-1 was incorporated into Balb/c T lymphocytes by incubation with Thy-1 at 200 μg/ml in 0.1 ml 9% PEG - 0.2% BSA - DAB at 20^{0}C for 30 min and washed prior to setting up in culture at 10^6 cells per 0.2 ml. In experiment 1, the cells were cultured without stimulation while in experiment 2, the cells were being stimulated to divide by cross-linked anti-Thy-1.2 MAbs and PMA. After the culture period, the cells were pelleted, and $[^{125}I]$F(ab')$_2$ MRC OX-7 was added for measurement of rat Thy-1. Rat thymocytes were cultured as a control.

Triggering via exogenous Thy-1

Triggering via the mouse Thy-1 was routinely very impressive, and experiments with BUdr labeling two days after culture showed that, essentially, all mouse T cells were being induced to divide (data not shown). In control experiments (not shown), 30H12 and MRC OX-7 MAbs strongly stimulated Thy-1.2 or Thy-1.1 cells, respectively, but showed no effect on the cells of opposite allotype. In contrast, triggering via the exogenous Thy-1 was generally unsuccessful (Table 4 shows one of 6 similar experiments). Further work is warranted, and it may be interesting to study biochemical events which mark triggering at early time points but the clear conclusion was that the incorporated Thy-1 did not mediate a functional effect in the same way as did the endogenous molecule.

Table 4. Triggering of Balb/c spleen T lymphocytes via Thy-1

| | cpm ($\times 10^{-3}$) 3H thymidine incorporated | | |
| | Culture Time (hr) | | |
Treatment	24	48	72
MRC OX-7	1.4	0.8	0.7
30H12	81.5	76.7	23.5
PMA	0.9	0.6	0.9

1×10^7 Balb/c T lymphocytes were treated with Thy-1 as in Table 3. The cells were washed and then incubated with MRC OX-7 or 30H12 MAbs (0.2 ml, 10 μg/ml) at 4°C for 1 h. The cells were then washed and cultured at 2×10^5 cells per well with RAM and PMA as described in Experimental Procedures.

COMMENTS

The experiments show that modest amounts of Thy-1 are incorporated into cells by culture with pure Thy-1 at high levels. Presumably, the micellar form of Thy-1 (Kuchel et al. 1978) does not readily expose the lipid groups, and the cell surface may also present a hydrophilic barrier to lipid incorporation. With PEG at levels which did not effect subsequent cell division, this incorporation was greatly stimulated.

Although rat brain Thy-1 could be stably incorporated into mouse T cells, this exogenous Thy-1 was ineffective as a mediator of mitogenesis. Two possible reasons for this failure can be suggested:

a) Some damage to the molecule has occurred during purification. The Thy-1 lipids have yet to be fully characterized, and if unsaturated lipids are present, these may undergo oxidation during purification.

b) The Thy-1 carbohydrate structures may be important in the mitogenic signaling process. The complex carbohydrates on rat brain Thy-1 will be very different to those on the endogenous Thy-1 of T lymphocytes (Parekh et al. 1987). Obviously, the experiments should be repeated using Thy-1.1 purified from mouse T lymphocytes.

REFERENCES

Campbell DG, Gagnon J, Reid KBM, Williams AF (1981) Biochem J 195:15-30

Hunt SV (1987) In: Klaus GGB (ed) Lymphocytes: A Practical Approach. IRL Press Oxford, pp 1-34

Kroczec RA, Gunter KC, Seligmann B, Shevach EM (1986) J Immunol 136:4379-4384

Kuchel PW, Campbell DG, Barclay AN, Williams AF (1978) Biochem J 169:411-417

Ledbetter JA, Herzenberg LA (1979) Immunol Revs 47:63-90

Mason DW, Williams AF (1980) Biochem J 187:1-20

Mason DW, Penhale WJ, Sedgwick JD (1987) In: Klaus GGB (ed) Lymphocytes: A Practical Approach. IRL Press Oxford, pp 35-54

Medof ME, Kinoshita T, Nussenzweig V (1984) J Exp Med 160:1558-1578

Parekh RB, Tse AGD, Dwek RA, Williams AF, Rademacher TW (1987) EMBO J 6:1233-1244

Tse AGD, Barclay AN, Watts A, Williams AF (1985) Science 230:1003-1008

Williams AF, Gagnon J (1982) Science 216:696-703

Identification and Characterization of Fatty Acid–Acylated Proteins in Cultured Cells by Radiolabeling

A.I. Magee

Laboratory of Cell Surface Interactions, National Institute for Medical Research, The Ridgeway, Mill Hill, London NW7 1AA, U.K.

INTRODUCTION

Since the first observation of the acylation of viral membrane proteins with long chain fatty acids in 1979 (Schmidt et al.), a large number of viral and normal cellular proteins have been found to be modified in this way (reviewed in Magee and Schlesinger 1982; Schmidt 1983; Sefton and Buss 1987). In general, two fairly well characterized acylation events have been observed: thioester linkage of mainly palmitic acid (C16:0) to cysteine residues, and amide linkage of exclusively myristic acid (C14:0) to N-terminal glycine residues (Magee and Courtneidge 1985; McIlhinney et al. 1985; Olson et al. 1985). In addition, there have been reports of oxyester-linked palmitate (Stoffel et al. 1983) and non-N-terminal amide-linked myristate (Kellie and Wrigglesworth 1987). Most recently, a number of ectoproteins of parasites and eukaryotic cells have been found to be associated with the plasma membrane via a complex glycosyl-phosphatidylinositol (G-PI) "tail" (Cross 1987). Many interesting proteins have now been identified as containing a covalently attached fatty acid or lipid; oncogene products such as pp60[src] and p21[ras] (Sefton et al. 1982), transmembrane proteins including receptors for transferrin (Omary and Trowbridge 1981) and insulin (Magee and Siddle 1986), and the cell adhesion molecule N-CAM (Hemperley et al. 1986). The function of covalently attached lipid in many cases is obscure, but it clearly is required for membrane association and function of p21[ras] (Willumsen et al. 1984) and pp60[src] (Kamps et al. 1986), and for cell surface association of G-PI-linked proteins (Low and Zilversmit 1980; Conzelman et al. 1986). Recently, the palmitoyl moiety of a number of proteins has been shown to turn over more rapidly than the polypeptide backbone, suggesting a dynamic role for this modification in protein function (Omary and Trowbridge 1981; Magee et al. 1987; Staufenbiel 1987). A few years ago, we set out to study the range of acyl modifications of proteins in eukaryotic cells, and developed a number of approaches to defining the association between fatty acids and proteins (Magee and Courtneidge 1985). In this chapter, I describe the current methods used in my laboratory.

EXPERIMENTAL PROCEDURES

Metabolic labeling of cultured cells with [^{3}H]fatty acids

a) Monolayers of cells in tissue culture dishes are grown to near confluence.

b) Cells are labeled usually for 4-6 h with 100 μCi/ml [9,10-^{3}H]palmitic acid (40-60 Ci/ mmol, TRK-760, Amersham International) or [9,10-^{3}H]myristic acid (10-30 Ci/mmol, NET 830, NEN, Darmstadt, FRG) in standard culture medium containing 1% foetal calf serum, 4 x the normal concentration of non-essential amino acid and 5 mM Na-pyruvate. The fatty acids are dried and maintained as stocks in ethanol at 10 mCi/ml at -20°C.

c) Cells are washed and lysed directly in gel-loading buffer containing dithiothreitol (Laemmli 1970), or in immunoprecipitation buffer (20 mM Tris, 150 mM NaCl, 1 mM EDTA, 0.5% (v/v) NP40, 0.5% (w/v) Na-deoxycholate, 0.1% (w/v) SDS, 1% (v/v) Trasylol (Sigma), 0.2 mM phenylmethylsulfonylfluoride, pH 7.4). Lysates are loaded directly, or after immunoprecipitation with an appropriate antibody, onto SDS-polyacrylamide gels (Laemmli 1970), followed by fluorography with 2,5-diphenyloxazole (PPO) in DMSO and exposure to preflashed Kodak X-AR5 film at -70°C (Laskey 1980). The dye front is run off the gel to remove free label and phospholipids.

Note

9,10-tritiated fatty acids are used to reduce reincorporation of label into other metabolic precursors and due to their high specific activity. ^{14}C-labeled fatty acids (particularly those labeled in the 1-position) should not be used due to susceptibility to degradation by β-oxidation to acetyl CoA and reincorporation of label which can give misleading results.

A number of labeling media have been assessed in this laboratory. The recipe given, gives the best results in our hands. Pyruvate is included to supply the acetyl CoA requirements of the cell and thus reduce reincorporation of label.

Relatively short labeling times (4-6 h) are usually chosen to reduce the problems of reincorporation of label. This varies greatly with cell type; some cells show specific labeling even after 24 h. A useful control is to run total cell lysates and compare the fatty acid-labeled samples with an [^{35}S]methionine-labeled sample. The labeling patterns with the two fatty acids should be substantially different from each other and from the [^{35}S]methionine-labeled proteins (Magee and Courtneidge 1985).

In our hands, fluorography with PPO-DMSO, although time consuming, gives the highest sensitivity. This is important for the detection of low amounts of tritium. Similarly, Kodak X-AR5 appears to be the most sensitive film available. Amersham Hyperfilm-MP is a good (and economical) substitute when maximum sensitivity is not required.

Linkage analysis of acylated proteins

This is most conveniently performed on duplicate slices from an SDS-polyacrylamide gel of [^{3}H]fatty acid-labeled and [^{35}S]methionine-labeled protein(s). The method can be adapted for protein in solution. Lanes are excised from a fresh gel and shaken at room temperature for 1 h with either:

a) 0.2 M KOH in methanol (this cleaves thio- and oxyesters but not amides) (Schmidt et al. 1979), methanol alone is used to treat a control strip; or

b) 1 M freshly prepared hydroxylamine hydrochloride titrated to pH 7.5 with NaOH (this cleaves thioesters, but not oxyesters or amides) (Magee et al. 1984). A duplicate lane is treated with 1 M Tris-Cl, pH 7.5, as a control.

The slices are then washed briefly three times with water for 5 min, and prepared for fluorography as above. This should not be performed on fixed or dried gels since both these procedures can lead to transacylation converting labile ester linkages to stable linkages of unknown nature (possibly amides).

Hydroxylamine treatment in solution can also be used as a mild method for removal of thioester-linked fatty acid for functional studies (Magee et al. 1987) or to prepare deacylated acceptor protein for *in vitro* acylation studies (Berger and Schmidt 1984).

Analysis of protein-bound label

Due to problems of interconversion of fatty acids via β-oxidation and chain elongation, and of reincorporation of label into other metabolic precursors, the protein-bound label derived from [^{3}H]fatty acids should be analyzed.

1. Procedure for soluble protein(s):
    Total protein-bound label can be analyzed as follows:

a) Precipitate labeled protein (e.g., cell lysate in 1% SDS) with 5 volumes of acetone containing 0.1 M HCl at -20°C for at least 1 h, and pellet the precipitate by centrifugation.

b) Dissolve the pellet in a minimum volume of 1% SDS, and reprecipitate twice as above.

c) Extract the pellet at least three times by trituration with chloroform - methanol (2:1) until no more free label is extracted into the organic solvent as determined by scintillation counting.

d) Extract the pellet finally with diethyl ether, and transfer it into a thick walled Pyrex tube. Dry under a nitrogen stream.

e) Evacuate the tube, hydrolyze with 6 N HCl at 110°C for 16 h, lyophilize twice.

f) Extract the residue twice with 0.5 ml hexane, and pool the extracts. Determine the radioactivity in the hexane extracts and in the residue (after dissolving in 1% SDS). Fatty acids should be quantitatively extracted into hexane, while label reincorporated into sugars and amino acids will be mainly in the hexane residue.

g) Evaporate the hexane extracts (e.g., in a Speed Vac), and dissolve in a small volume (2-5 μl) of chloroform - methanol (2:1). Spot onto a RP18 thin-layer plate (Merck 13724), and develop in acetonitrile - acetic acid (90:10). Spot authentic [^{3}H]fatty acids in parallel lanes as markers.

h) Dry the plate and detect the radioactivity by spraying with En^{3}hance spray (NEN) and exposure to preflashed Kodak X-AR5 film at -70°C, or by scraping 1 cm lengths of adsorbent and scintillation counting.

2. Procedure for gel bands:

a) Following electrophoresis, locate the band of interest either by using molecular weight standards, or by fluorography using Na-salicylate (Chamberlain 1979). PPO-treated gels cannot be used.

b) Excise the band from the wet or dried gel, and wash three times with 0.5 ml water for 5 min each wash (agitate), during which time the dried gel piece rehydrates and the salicylate is washed out.

c) Transfer the gel piece to a hydrolysis tube, lyophilize, and proceed as for total protein (step e) above. Alternatively, protein can be digested out of the gel with 200 μg/ml pronase (Calbiochem) in 20 mM ammonium bicarbonate, 0.05% SDS, pH 8.0, at 37°C for 48 h, followed by lyophilization.

COMMENTS

The techniques described should allow the detection of acylated proteins, the determination of the linkage of fatty acid to protein, and the identification of the incorporated label. All the methods described can be easily performed with routine laboratory equipment and procedures. The identification of new acyl proteins of known function will greatly aid in the assessment of the physiological role(s) of covalently attached fatty acid.

Acknowledgments. The work in my laboratory is supported by the Medical Research Council. I thank Lourdes Gutierrez and Julie Childs for assistance, and Marilyn Brennan for typing the manuscript.

REFERENCES

Berger M, Schmidt MFG (1984) J Biol Chem 259:7245-7252

Chamberlain JP (1979) Anal Biochem 98:132-135

Conzelman A, Spiazzi A, Hyman R, Bron C (1986) EMBO J 5:3291-3296

Cross GAM (1987) Cell 48:179-181

Hemperley JJ, Edelman GM, Cunningham BA (1986) Proc Natl Acad Sci USA 83:9822-9826

Kamps MP, Buss JE, Sefton BM (1986) Cell 45:105-112

Kellie S, Wrigglesworth NM (1987) FEBS Lett 213:428-432

Laemmli UK (1970) Nature 227:680-685

Laskey RA (1980) In: Grossman L, Moldave K (eds) Methods in Enzymology, Vol 65, Academic Press New York London, pp 363-371

Low MG, Zilversmit DB (1980) Biochemistry 19:3913-3918

Magee AI, Schlesinger MJ (1982) Biochim Biophys Acta 694:279-289

Magee AI, Koyama AH, Malfer C, Wen D, Schlesinger MJ (1984) Biochim Biophys Acta 798: 156-166

Magee AI, Courtneidge SA (1985) EMBO J 4:1137-1144

Magee AI, Siddle K (1986) Biochem Soc Trans 14:1103-1104

Magee AI, Gutierrez L, McKay IA, Marshall CJ, Hall A (1987) EMBO J (in press)

McIlhinney RAJ, Pelly SJ, Chadwick JK, Cowley GP (1985) EMBO J 4:1145-1152

Olson EN, Towler DA, Glaser L (1985) J Biol Chem 260:3784-3790

Omary MB, Trowbridge IS (1981) J Biol Chem 256:12888-12892

Schmidt MFG, Bracha M, Schlesinger MJ (1979) Proc Natl Acad Sci USA 76:1687-1691

Schmidt MFG, (1983) Current Topics Microbiol Immunol 102:101-129

Sefton BM, Trowbridge IS, Cooper JA, Scolnick EM (1982) Cell 31:465-474

Sefton BM, Buss JE (1987) J Cell Biol 107:1449-1453

Staufenbiel M (1987) Mol Cell Biol 7:2981-2984

Stoffel W, Hillen H, Schroder W, Deutzmann R (1983) Hoppe-Seyler's Z Physiol Chem 364: 1455-1466

Willumsen BM, Norris K, Papageorge AG, Hubbert NL, Lowy DR (1984) EMBO J 3:2581-2585

Chemical Analysis of Fatty Acids Covalently Attached to Proteins

P.J. Casey* and J.E. Buss[+]

*Department of Pharmacology, Southwestern Graduate School of Biomedical Sciences,
The University of Texas Health Science Center, 5323 Harry Hines Boulevard, Dallas,
Texas 75235-9041, USA
[+]La Jolla Cancer Research Foundation, Cancer Research Center (NCI),
10901 North Torrey Pines Road, La Jolla, CA 92037, USA

INTRODUCTION

The list of proteins with a fatty acyl group attached directly to the peptide backbone is an interesting melange of enzymes and structural proteins which play important but diverse roles in cellular metabolism and transformation as well as virion architecture (Table 1) (Sefton and Buss 1987).

For many of these proteins, metabolic labeling with [^{3}H]fatty acids and immunoprecipitation has been the technique used for characterization of the attached acyl group. As informative as this approach has been, it has some limitations. First, it depends upon the availability of an antiserum of sufficiently high titer in order to be useful for immunoprecipitations. Second, the incorporation of [^{3}H]label may be misleading as one fatty acid may be metabolized to another, particularly during prolonged labeling periods. Third, there is the worry that failure to incorporate detectable amounts of [^{3}H]palmitic acid may be due to dilution of the label in the large intracellular pool of palmitate. As additional proteins are characterized, new methods for detection and identification of covalently attached fatty acids will be needed.

We have been studying the guanine nucleotide binding regulatory proteins (G proteins) G_s, G_i, G_o, and transducin (Gilman 1987). In addition to being important regulators of the activity of adenyl cyclase ($G_{s\alpha}$ and $G_{i\alpha}$) and retinal phosphodiesterase (transducin), these proteins are related to another GTP-binding protein, p21^{c-ras} (Hurley et al. 1984). p21^{c-ras} is acylated via attachment of palmitic acid to a cysteine residue near the carboxy terminus of protein (Chen et al. 1985). We began these studies to determine whether acylation was a feature of other G-proteins.

Immunoprecipitation experiments using human astrocytoma cells labeled metabolically with either [^{3}H]myristic acid or [^{3}H]palmitic acid revealed that $G_{i\alpha}$ incorporated [^{3}H]myristic acid but not [^{3}H]palmitic acid, whereas $G_{s\alpha}$ failed to become labeled with either fatty acid (Buss et al. 1987). This method of direct chemical analysis of covalently attached fatty acids was developed to confirm the presence of myristic acid on $G_{i\alpha}$, and to search for fatty acyl modifications of other G-protein subunits.

Table 1

Myristylated proteins	Function / Activity
p60^{src}	protein kinase
p56^{lck}	protein kinase
Catalytic subunit cAMP-dependent protein kinase	protein kinase
Calcineurin B	protein phosphatase
NADH-cytochrome b$_5$ reductase	electron transfer protein
p15^{gag} of mammalian leukemia viruses	virion structural protein
p120$^{gag-abl}$, p85$^{gag-fes}$, p29$^{gag-ras}$	transforming proteins
Poliovirus VP4	virion capsid protein
Hepatitis B surface antigen	virion surface antigen
$G_{i\alpha}$, $G_{o\alpha}$	GTP-binding proteins

Palmitylated proteins	Function / Activity
p21^{c-ras}	GTP-binding protein
Mammalian transferrin receptor	cell surface receptor
HLA B glycoprotein	histocompatability antigen
Apolipoprotein A-1	serum protein
Ankyrin	cytoskeletal protein
SV40 large T antigen	tumor antigen
Vesicular stomatitis virus glycoprotein	virion envelope protein
Alphavirus E2 glycoprotein	virion envelope protein
Parainfluenza virus glycoprotein	virion envelope protein

EXPERIMENTAL PROCEDURES

Materials

1. High performance liquid chromatography: Beckman Model 341 or equivalent, including a UV detector capable of measuring changes in absorbance at 254 nm of 0.002 units. The column used is a 0.45 x 25 cm Altex Ultrasphere-octyl 5 μ or equivalent.
2. Clinical centrifuge: IEC Model HN-S or equivalent
3. Polyacrylamide gel electrophoresis: standard equipment
4. Nitrogen dryer
5. Glass tubes: 13 x 100 mm with screw-on Teflon lined caps
6. Solvents: all HPLC grade; H_2O, MeOH, $CHCl_3$, CH_3CN
7. Hydrolysis reagents: 6 N HCl, sequencing grade (Pierce 24309 or equivalent); NaOH, highest purity obtainable
8. Derivatizing reagents: α,p-dibromoacetophenone and dicylohexyl-18-crown-6, available in a kit from Applied Science (#18036)
9. C_{18} Sep-Pak cartridges: Waters #351910 or equivalent
10. Heptadecanoic acid: Analabs #LSA 039 or equivalent

Methods

1. <u>Sample preparation</u>

The protein to be analyzed (1-2 nmol) is prepared in standard Laemmli sample buffer (Laemmli 1970) in a final volume of 300 μl. If desired, the protein can be precipitated first by adding 7 volumes of ice-cold acetone, incubating at 0°C for 3 h, and centrifuged. The precipitated sample is then dissolved in 300 μl of sample buffer. The proteins of interest are resolved by electrophoresis through 12% polyacrylamide gels (Laemmli 1970). This procedure has the dual advantage of stripping the protein of non-covalently linked lipids, and packaging the protein into a convenient form (a gel slice) for subsequent manipulations.

The proteins are located after electrophoresis by staining the gel for 5 min with 0.25% Coomassie blue in 30% isopropanol - 10% acetic acid (see *Notes*). The gel is destained for 16 h with three changes of 30% MeOH - 10% acetic acid, and then for 4 h with two changes of 50% MeOH. From this point on, all solvents used are HPLC grade, and all materials which come into contact with the gel slice or extracts derived therefrom, are rinsed with $CHCl_3$ - MeOH (1:2) prior to their use (see *Notes*).

The protein bands are now excised from the gel, placed in 13 x 100 mm test tubes, and washed for 16 h with three changes of 50% MeOH. It is advisable to excise equivalent slices from a lane in which only the sample buffer was loaded, and to carry these samples through the remainder of the procedure to determine the background levels of fatty acids picked up during the processing. After the final wash, the solvent is removed, and the gel slices dried under a stream of nitrogen.

2. Hydrolysis

The gel slices are suspended in 0.7 ml of 1.5 N NaOH and incubated in a waterbath at $30^{\circ}C$ for 3 h. This treatment is sufficient to hydrolyze thioester- and ester-linked fatty acids from the protein. The solution containing the gel slice is then acidified to pH 1-2 with 6 N HCl (~150 μl), and any released fatty acids are extracted with $CHCl_3$/MeOH by the method of Bligh and Dyer (1959). Briefly, 3.0 ml of $CHCl_3$ - MeOH (1:2) is added to the sample, and the mixture is incubated for 10 min with occasional vortexing. The fatty acids are separated from the water soluble material and the gel slice containing the protein by diluting the extraction mixture with 0.8 ml of $CHCl_3$ followed by 0.8 ml of H_2O. This mixture is centrifuged in the clinical centrifuge at 1000 rpm at room temperature for 2 min. The $CHCl_3$ (lower) phase is then removed by inserting a pipette through the MeOH/H_2O phase. An additional 0.8 ml of $CHCl_3$ is added to the mixture containing the gel slice, the sample incubated for 5 min with occasional vortexing, and then centrifuged and the $CHCl_3$ layer removed as above. The combined $CHCl_3$ fractions are placed in a clean 13 x 100 mm test tube, dried under a stream of nitrogen, and dissolved in 1.5 ml of CH_3CN for derivatization.

The gel slices in residual $CHCl_3$/MeOH are dried under a stream of nitrogen, suspended in 0.7 ml of 6 N HCl, and incubated at $100^{\circ}C$ for 4 h. This treatment hydrolyzes amide-linked fatty acids from the protein. Following this treatment, the released fatty acids are extracted from the mixture as described above except that the volume of $CHCl_3$/MeOH initially added is 2.6 ml, and the $CHCl_3$ and H_2O additions are 0.7 ml each. Again, the combined $CHCl_3$ phases are dried under nitrogen and dissolved in 1.5 ml of CH_3CN.

3. Derivatization

The extracted fatty acids are derivatized with dibromoacetophenone in the presence of a crown ether catalyst by a modification of the method of Durst et al. (1975). To the base- or acid-hydrolyzed extracts in 1.5 ml of CH_3CN, 5 μl of 200 μM heptadecanoic acid in CH_3CN are added. This fatty acid which is essentially unknown in nature, is added to monitor recovery during derivatization and chromatography. To each tube, 40-60 mg of $KHCO_3$ and the Applied Science derivatizing reagents are then added. These two reagents, α,p-dibromo-acetophenone (phenone) and dicyclohexyl-18-crown-6 (crown ether), are supplied in sealed ampules containing 0.2 mmol of each reagent in 1 ml of CH_3CN. Working stocks of these two reagents are prepared by making 16-fold dilutions of each in CH_3CN. The working stocks are stored under nitrogen at $2^{\circ}C$ and are stable for several weeks. A detailed description of the chemistry involved in the derivatization including a list of features and applications of the process is available from the manufacturer (Applied Science, Dearfield, IL).

To the samples containing heptadecanoic acid and $KHCO_3$ in CH_3CN, 25 μl of the phenone and 40 μl of the crown ether working stock solutions are added. A stream of nitrogen is passed into each tube which are then capped tightly and heated (preferably on a heating block) at $85^\circ C$ for 45 min with occasional vortexing. The samples are then removed and allowed to cool to room temperature. It is important that samples and reagents be kept dry. Nitrogen is the easiest method, and all bottles or tubes should be flushed before storage.

The derivatized and non-derivatized fatty acids are separated by filtration (chromatography) through C_{18} Sep-Pak cartridges. The cartridge is fitted with a 6 ml disposable syringe, and first washed with 5 ml each of $CHCl_3$ - MeOH (1:2) and CH_3CN by pouring the solvents into the syringe and forcing them through the Sep-Pak with the syringe plunger. Using the same procedure, the sample is poured into the Sep-Pak, forced through into a clean 13 x 100 mm tube, and the sample tube is rinsed with 2 ml of CH_3CN, and this sample also pushed through the Sep-Pak into the tube containing the first flow-through. The derivatized fatty acids do not bind to the C_{18} resin in CH_3CN while the underivatized do bind under these conditions and can be removed by washing the cartridge three times with 1.5 ml of $CHCl_3$ - MeOH (1:2). The cartridge is then rinsed twice with 2.5 ml of CH_3CN, and another sample is processed. A single Sep-Pak can be used to process up to 15 samples if regenerated with $CHCl_3$/MeOH after each. The derivatized samples in CH_3CN are dried under a stream of nitrogen, and dissolved in 100 μl of either MeOH - H_2O (88:12) if the samples are to be chromatographed on the HPLC immediately, or CH_3CN if the samples are to be stored (see *Notes*) for any period of time before chromatography. The derivatized fatty acids are somewhat unstable in the presence of any H_2O. It is advisable, therefore, to process them through the HPLC as they are prepared.

4. HPLC of derivatized samples

The chromatography of derivatized samples is performed isocractically using MeOH - H_2O (88:12) as the mobile phase and monitoring absorbance of 254 nm (0.05 AUFS). The flow rate is 1.0 ml/min. The elution times of the derivatized fatty acids are very sensitive to the MeOH concentration; it is advisable, therefore, to titrate the mobile phase with either H_2O or MeOH such that the elution time of derivatized heptadecanoate is ~40 min. This results in approximate elution times for derivatized fatty acids as follows: myristate 20 min; palmitate 31 min; stearate 50 min. The column is standardized by chromatographing any desired fatty acid or mixture of fatty acids which have been dissolved in CH_3CN and derivatized as described in section 3. The derivatized base and acid extracts are similarly chromatographed; injections can be performed at 60 min intervals. If the procedure is being performed with radioactive labeled fatty acids, or if further characterization of the eluted material is desired, the peaks can be collected, dried under nitrogen, and stored at -20°C in CH_3CN. The minimum stoichiometry of acylation can be estimated directly from the HPLC analysis. The peak areas from the known amount (1 nmol) of the added heptadecanoate standard and the experimental peak(s) are compared to determine the amount of fatty acid released from the known amount of protein used initially. This value of moles of fatty acid per mole of protein will be an underestimate as losses during gel electrophoresis, hydrolysis, and extraction are not included.

RESULTS

The accompanying figure shows the results obtained from the processing of approximately 2 nmol of two G-proteins by this procedure (Buss et al. 1987). The *alpha* subunits of at least two members of this family (designated by their apparent molecular weights as α_{39} ($G_{o\alpha}$) and α_{41} ($G_{i\alpha}$)) contain acid-labile (presumably amide-linked) myristic acid which can be readily detected by this procedure (upper and center panels). The small peaks of absorbance comigrating with derivatized standards of palmitate, oleate, and stearate, are apparently due to trace contamination by these fatty acids of the materials used in the processing as they are also observed in the control gel slice (lower panel). Myristate was not detected in transducin or $G_{s\alpha}$.

No base-labile (thio- or oxyester-linked) fatty acids were observed on any of these G-proteins (Buss et al. 1987). It should be noted that we have not yet processed a protein known to contain a base-labile fatty acid by this procedure. Therefore, the method has been verified to detect only acid-labile (amide-linked) fatty acids at this time. However, we feel that base-labile fatty acids can also be detected by this method as the sample processing contains no steps which should result in removal of this class of covalently linked fatty acids prior to the base hydrolysis step.

COMMENTS

This method of gel electrophoresis, hydrolysis *in situ*, and derivatization of released lipids allows fatty acids covalently linked to proteins to be detected and quantified with equipment available in most laboratories. Previous work had relied upon gas chromatographic (GC)-, direct chemical ionization (DCI)-, or fast atom bombardment (FAB)-mass spectrometry (Carr et al. 1982), all of which require specialized instruments and rigorously purified pro-

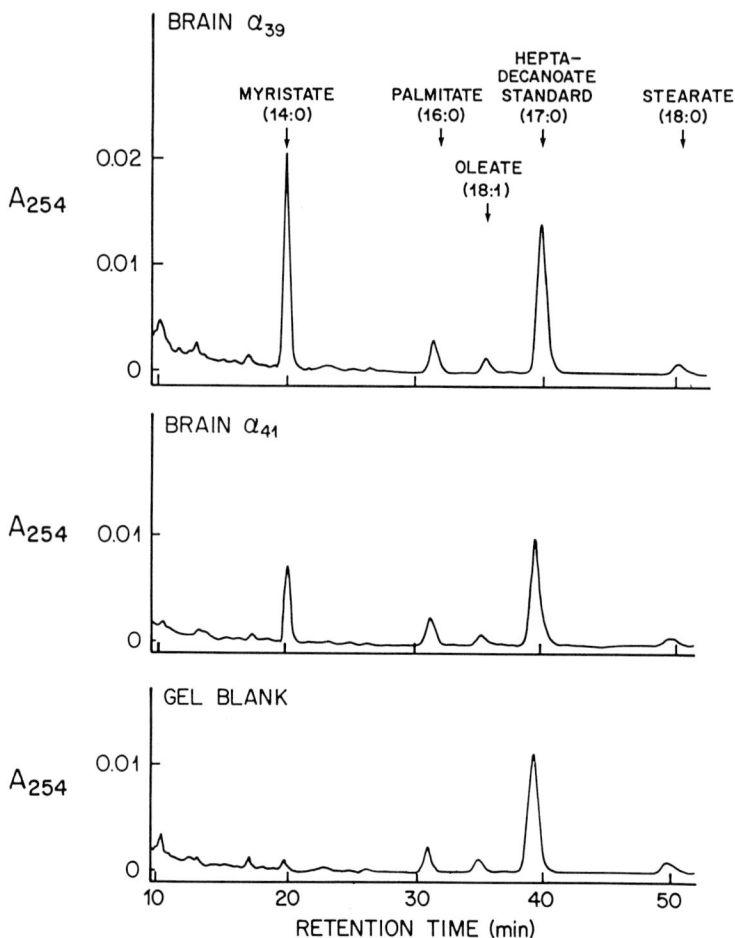

Fig. 1. Chemical analysis of acid-labile fatty acids on G-protein α subunits. The chloroform/methanol extracts of gel slices containing α_{39} (upper panel), α_{41} (center panel), or the buffer lane (gel blank, lower panel) were subjected to derivatization and chromatography as described under "Methods". The initial quantities of each subunit loaded on the gel were 90 μg of α_{39} and 60 μg of α_{41}. The arrows identify the elution positions of the appropriate fatty acid standards derivatized by the same procedure as the samples. The experiment shown is representative of three separate experiments using two different preparations of each protein. (Reprinted from Buss et al. 1987)

teins or peptides. Although, this derivatization method is a very sensitive technique for detecting fatty acids which have been released from a protein, it still requires a sizable amount of protein (1-2 nmol). The biggest problem is the presence of lipids on glass-ware or plastics (including the investigator's gloves (see *Notes*)) which give spurious peaks in the HPLC analysis. A major advantage of this procedure is its adaptation to gel-purified protein so that if contaminants or other subunits can be separated by any electrophoretic technique, partially purified protein can be analyzed without further preparation. Electrophoresis should be used even for highly purified proteins as it removes contaminating lipids wich frequently fail to be removed during conventional purification steps, or which may be added inadvertently with "fatty acid free" BSA. The preliminary steps of base and acid hydrolysis of the gel slice are also useful for radiolabeled proteins. Once, the [^{3}H]labeled fatty acids have been extracted into solvent, they can be analyzed directly by thin layer chromatography and fluorography (Buss and Sefton 1985).

Notes

1. All solvent mixtures are v/v.
2. Disposable gloves should be worn at all times to prevent contamination of the samples or glass-ware with lipids. Cotton photography gloves can also be used.
3. Fatty acids (both derivatized and underivatized) are routinely stored in CH_3CN under nitrogen at $-20^\circ C$.

Acknowledgments. This work was supported by National Institutes of Health Grant GM 34497 and American Cancer Society Grant BC5551 to Alfred G. Gilman, and the Raymond and Ellen Willie Chair in Molecular Neuropharmacology. We thank Dr. Robert Munford and Alice Erwin for their advice on this procedure, Dr. Susanne Mumby for unfailing encouragement, and Dr. Alfred G. Gilman for guidance and support.

REFERENCES

Applied Science report #18036 (Applied Science, Dearfield, IL)
Bligh EG, Dyer WJ (1959) Can J Biochem Physiol 37:911-917
Buss JE, Sefton BM (1985) J Virol 53:7-12
Buss JE, Mumby SM, Casey PJ, Gilman AG, Sefton BM (1987) Proc Natl Acad Sci USA 84: 7493- 7497
Carr SA, Biemann K, Shoji S, Parmelee DC, Titani K (1982) Proc Natl Acad Sci USA 79: 6128-6131
Chen ZQ, Ulsh LS, Du Bois G, Shih TY (1985) J Virol 56:607-612
Durst HD, Milano M, Kikta EJ, Connelly SA, Grushka E (1975) Anal Chem 47:1797-1801
Gilman AG (1987) Ann Rev Biochem 56:615-549
Hurley JB, Simon MI, Teplow DB, Robishaw JD, Gilman AG (1984) Science 226:860-862
Laemmli UK (1970) Nature 227:680-685
Sefton BM, Buss JE (1987) J Cell Biol 104:1449-1453

Identification and Characterization of Proteins Undergoing Reversible Fatty Acid Acylation

M. Staufenbiel

Max-Planck-Institut für Zellbiologie, Abt. Traub, Rosenhof,
D-6802 Ladenburg bei Heidelberg, Federal Republic of Germany

INTRODUCTION

In eukaryotic cells, two classes of fatty acid acylated proteins exist containing either amide-linked myristic acid, or carrying predominantly palmitic acid bound via thioester or ester linkages (Magee and Courtneidge 1985; McIlhinney et al. 1985; Olson et al. 1985). Modification by myristic acid is tightly coupled to protein synthesis (Magee and Courtneidge 1985; McIlhinney et al. 1985; Olson et al. 1985), and is apparently quite stable (Buss et al. 1984). Palmitylation also seems to occur during the biogenesis of proteins (Schmidt and Schlesinger 1980), however, it has been shown recently that the palmityl moiety of some proteins undergoes a sometimes rapid turnover (Omary and Trowbridge 1981; Magee et al. 1987; Staufenbiel 1987; Staufenbiel, submitted for publication). With erythrocyte acyl proteins, differences were found between the half-lives of their palmityl groups suggesting an independent fatty acid turnover within the same cell type. Deacylation of these proteins was not due to a chemically labile bond between polypeptide and fatty acid, rather it appeared to be physiologically induced and may require enzyme catalysis (Staufenbiel, submitted for publication). Together, these results show that dynamic palmitylation has a possible regulatory significance and may serve to modulate diverse functions of proteins. Reversible palmitylation has been shown to occur at the plasma membrane and its associated cytoskeleton (Staufenbiel, submitted for publication), but it remains to be determined whether the palmitic acid modification at intracellular membranes (Schmidt and Schlesinger 1980) can also be reversed. In this chapter, I describe the procedure currently used to efficiently label cells with fatty acids and to do pulse-chase experiments. The techniques were developed and characterized using erythrocytes, but they can be applied to cultured cells as well.

EXPERIMENTAL PROCEDURES

Labeling of cells with radioactive fatty acid and analysis of proteins and lipids

Radioactive fatty acids are usually supplied in ethanol or toluene. The desired amount is pipetted into a glass tube using gas-tight syringes (Hamilton) for exact pipetting. A disposable tissue is placed over the tube opening to trap aerosols, and the solvent is removed by evaporation under vacuum (water pump). Thereafter, labeling medium is added which contains bovine albumin (fatty acid free; Sigma No. A-7030). During the following incubation at $37^{o}C$ to warm up the medium (about 15 min), almost all fatty acid binds non-covalently to albumin. Then, in a small volume of medium, the cells are added if labeling is done in suspension, and the tube is incubated on a rocker at $37^{o}C$. For cells grown on a substratum, the labeling medium is transferred to a petri dish, and the incubation proceeds as usual. Note that no organic solvents are present in the labeling medium if this procedure is followed.

To obtain strong labeling, about 1.5 x 10^8 erythrocytes (5 x 10^6 cultured suspension cells, or a near confluent monolayer on a 6 cm dish) are incubated in 1 ml of labeling medium containing 100 μCi [^{3}H]fatty acid (e.g., [9,10-^{3}H]palmitic acid, [9,10-^{3}H]myristic acid (New England Nuclear, Amersham); use the highest specific activity available). The molar ratio of fatty acid to albumin in the labeling medium should be between 1 and 2. If this ratio is lower, too much fatty acid remains bound to albumin in the medium at the end of the labeling period. If no albumin is present in glass tubes, most of the fatty acid rapidly associates with the cells, but labeling of proteins is not, or only slightly, better than with albumin. However, in the absence of albumin, a large portion of fatty acid may bind to plastic materials (tubes, petri dishes, pipette tips, etc.), while little label is lost if albumin-bound fatty acid comes in contact with plastic. Therefore, it is advisable to use albumin. We use Dulbecco modified Eagle medium (DMEM) or other media for labeling, but try to leave out serum to exclude unlabeled fatty acids. If this cannot be avoided, the smallest possible amount of foetal calf serum is added. For labeling of cells which actively synthesize proteins, the medium is supplemented with 5 mM sodium pyruvate and 3 times the DMEM concentration of non-essential amino acids to minimize degradation of fatty acids and conversion of label into amino acids. The optimum labeling time depends very much on the protein under study and its possible fatty acid turnover rate, but normally a 6 h period is suitable for most proteins.

Labeled proteins and lipids are analyzed by standard techniques. Polypeptides are dissolved in SDS-sample buffer (containing 25 mM DTT; larger amounts of reducing agents in combination with long boiling times may result in the loss of fatty acids) and separated by SDS-polyacrylamide gel electrophoresis, stained with Coomassie brilliant blue (in e.g., 25% isopropanol, 10% acetic acid), and destained with 10% acetic acid without loss of fatty acids. Radioactive bands are visualized by fluorography using the DMSO-PPO method (Bonner and

Laskey 1974) and Kodak X-Omat AR5 films. In our hands, this is most sensitive. The exposure times required vary considerably; to start with, a 1 week period is suggested. To ensure
that the label is covalently bound, chloroform/methanol extracted proteins (see below)
should also be analyzed. Thioester- or ester-linked fatty acids can be removed by treatment
with 0.1 M KOH in methanol indicating that the label has not been incorporated into the
polypeptide backbone. Both treatments can also be done on polyacrylamide gels after separation of polypeptides. The treatments are at room temperature for 2 h, after which gels have
to be rehydrated by washing in water for 1 h before they can be prepared for fluorography.
Phospholipids are isolated by chloroform - methanol (2:1; v/v) extraction (3 x a 100-fold
sample volume) concentrated by evaporation of solvent, and separated by thin-layer chromatography on silica gel plates developed in chloroform - methanol - acetic acid - water
(52.5:50:4:1; v/v) (Schimmel and Traub 1987). Marker phospholipids are visualized with
iodine vapor. The radioactivity in the different spots can be determined directly with a
thin-layer chromatographic linear analyzer and integrator (Berthold LB 282). The plates can
also be prepared for fluorography with PPO-methyl-naphthalin (Bonner and Stedman 1978).
By aligning the X-ray film over the chromatogram, radioactive spots can be located,
scratched out, and counted in a standard liquid scintillation counter.

Pulse-chase experiments with fatty acid

Pulse-labeling is performed as described above in medium lacking pyruvate or additional
amino acids, but with albumin (molar ratio of fatty acid to albumin, 2). About 4 μCi of label
is added per 1 x 10^7 erythrocytes or 50 μCi per 1 x 10^7 cultured cells. With more fatty acid,
the chase may be inefficient or even impossible (steady increase in labeling). About 1.5 x 10^9
erythrocytes (or 5 x 10^7 suspension-cultured cells) are incubated in 1 ml of labeling medium
at 37°C for 10 min. In the next step, cells are washed 7 to 10 times in medium (e.g., DMEM)
containing 5 mg albumin/ml, thereby, removing more than 95% of the fatty acid which has-
not been incorporated into proteins or phospholipids. This latter step is more efficient if the
[^{3}H]fatty acid was bound to albumin during the labeling period. Thereafter, the cells are suspended in medium lacking albumin (same cell concentration as used for pulse-labeling), and
transferred to a glass tube containing solid unlabeled palmitic acid (250 nmol palmitic acid
in ethanol; solvent evaporated). Quick loading of cells with unlabeled fatty acid is achieved
in this way since more than 90% of the fatty acid associates with the cells during incubation
at 37°C for 3-5 min. Cells on petri dishes are incubated for 5 min in medium containing 10
mg albumin and 2.5 mmol palmitic acid/ml. Following incubation, cells are chased in medium containing 1 mg albumin/ml, 100 μM palmitic acid and 5% foetal calf serum. Palmitic
acid can often be omitted from this step since little is taken up and the efficiency of the

chase is not improved by its presence. For cells active in protein synthesis, sodium pyruvate and non-essential amino acids are added to all media used after pulse-labeling. Since washing requires some time, samples should be taken already during washing and during the chase. The samples are then analyzed as described above. To detect labeled polypeptides, exposure times of about 4-6 weeks will be required. Fluorographs within the linear relationship of radioactivity and absorbance of image (determined by quantifying serially diluted samples) are scanned with a densitometer, and the peak areas integrated. It is difficult to determine the radioactivity directly since the amount of [^{3}H]fatty acid incorporated during a short pulse is usually too small. Most of the [^{3}H]palmitic acid is incorporated into the cellular phospholipids during pulse-labeling where it undergoes turnover. This huge pool cannot be diluted appropriately without lysing the cells. Unlabeled palmitic acid added at the beginning of the chase, rapidly enters the precursor pool(s) for protein and phospholipid acylation (palmityl-coenzyme A probably), and sufficiently dilutes the small pool(s). This dilution effect is lost as the unlabeled fatty acid is incorporated, and labeled fatty acid which is set free, mostly from the phospholipids, is reused. Therefore, the chase becomes less and less efficient after longer periods of time. Kinetics of fatty acid turnover on proteins are often biphasic which may reflect a reuse of labeled fatty acid rather than subpopulations of the acyl proteins. The half-lives of fatty acids can be best determined from the initial turnover rates. They should be compared with the half-lives of the polypeptide backbone.

REFERENCES

Bonner WM, Laskey RA (1974) Eur J Biochem 46:83-88
Bonner WM, Stedman JD (1978) Anal Biochem 89:247-256
Buss JE, Kamps MP, Sefton BM (1984) Mol Cell Biol 4:2697-2704
Magee AI, Courtneidge SA (1985) EMBO J 4:1137-1144
Magee AI, Gutierrez L, McKay IA, Marshall CJ, Hall A (1987) EMBO J 6:3353-3357
McIlhinney RAJ, Pelly SJ, Chadwick JK, Cowley GP (1985) EMBO J 4:1145-1152
Olson EN, Towler DA, Glaser L (1985) J Biol Chem 260:3784-3790
Omary MB, Trowbridge IS (1981) J Biol Chem 256:12888-12892
Schimmel H, Traub P (1987) Lipids 22:95-103
Schmidt MFG, Schlesinger MJ (1980) J Biol Chem 255:3334-3339
Staufenbiel M (1987) Mol Cell Biol 7:2981-2984
Staufenbiel M (submitted for publication)

Acylation of Proteins in Isolated Mitochondria

E. Sigel, L. Lehmann, and J.W. Stucki

Pharmakologisches Institut, Universität Bern, Friedbühlstrasse 49, CH-3010 Bern, Switzerland

INTRODUCTION

Fatty acids are known to be regulators of several mitochondrial functions such as the ATPase, respiratory enzymes, or carrier mediated transport accross the inner membrane (Wojtczak 1976; Batayneh et al. 1986). The covalent modification of proteins by fatty acids occurring either co-translationally with myristate, or post-translationally with palmitate, has gained wide attention (for review see Sefton and Buss 1987 and references therein). Protein acylation has been implicated in the provision of proteins with a hydrophobic anker for membrane binding, but its function is largely unclear. We had decided to investigate whether such a covalent modification can occur in isolated, highly purified mitochondria. To our surprise, we found such reactions to occur, but they differ from cytoplasmic acylation in chain-length specificity (Stucki et al. 1988, submitted for publication). In this chapter, we describe how this mitochondrial acylation can be demonstrated (Experiment A), how the fatty acid chain-length specificity of this process may be shown (Experiment B), and how the chemical nature of the bound radioactivity may be determined (Experiment C).

MATERIALS

Liver mitochondria

Liver mitochondria were prepared from 250 g male rats according to the procedures of Johnson and Lardy (1967) except that during washing, bovine serum albumin (1 mg/ml) was included in the medium in order to remove a possible contamination by free fatty acids. At the end of the preparation, the mitochondria were suspended in a small volume, and the protein concentration was quickly determined as follows according to a variant of the biuret reaction. The sample containing 0.1-0.5 mg protein was diluted with distilled water to 0.2 ml. Then, 0.7 ml of 9% NaOH - 0.4% Na-cholate were added. After mixing, 0.1 ml of 1% $CuSO_4$

was added, and the mixture centrifuged in a microfuge for 1 min. After 5 min, the absorbance was determined at 540 nm, and compared to values obtained with a bovine serum albumin standard.

Where indicated, the isolated mitochondria were further treated with 0.2% digitonin in order to remove the outer mitochondrial membrane and contaminating membrane fractions (Schnaitmann et al. 1961) with the modifications described in procedure XIIIA1 in Peterson et al. (1978). Glutamate dehydrogenase (Schmidt 1970), arylsulfatase (Baum et al. 1959), glucose-6-phosphatase (Baginski et al. 1970), and catalase (Aebi 1970) activity were taken as markers for mitochondrial matrix, and for lysosomal, endoplasmic reticulum, peroxisomal contaminations, respectively. Contamination by lysosomes may be reduced about 10-fold, by endoplasmic reticulum about 5-fold, and by peroxisomes about 3-fold. Mitochondria and mitoplasts should be prepared in the shortest time possible and used quickly after isolation.

Experiments A and B

1.5 ml Eppendorf tubes
Thermostated waterbath at 37°C
Heating block at 95°C
Equipment for SDS-gel electrophoresis and autoradiography
58.0 mCi/mmol [1-^{14}C]myristic acid (New England Nuclear)
Myristic acid
1,3-butanediol
Incubation medium for the acylation experiment:
60 mM KCl, 7.5 mM K-phosphate, pH 7.4, 40 mM triethanolamine hydrochloride
(neutralized with KOH), 15 mM K-succinate, 2 mM K-glutamate, 2 mM K-malate,
1 mM K-ATP, 1 mM $MgCl_2$, 0.65 μg/ml rotenone
Sample buffer for gel electrophoresis, 3 x concentrated:
30 mM Na-phosphate, pH 7.0, 7.5% SDS (w/v), 30% glycerol (v/v),
10 mM dithiothreitol, 0.5 mg/ml bromphenol blue

Experiment C

En^{3}hance spray (New England Nuclear)
Methanol, chloroform
Thin-layer chromatography plates (C_{18}-silanized silicagel, Sigma)

Mobile phase for thin-layer chromatography:

Methanol - dioxan - chloroform - 50 mM glycine, pH 3.0 in water (4:3:2:3)

Hydrolyzing solutions:

1 M KOH in 20% methanol in water (v/v)

1 M hydroxylamine, pH 9.9

1 M Tris-HCl, pH 9.9

EXPERIMENTAL PROCEDURES

Experiment A:
Acylation of mitochondria and mitoplasts

The radioactive myristic acid can be obtained in ethanol. For the experiment, a suitable volume of the acid solution to give a final concentration of 75 μM is pipetted into an Eppendorf tube, the ethanol is evaporated under a stream of nitrogen gas, and the acid redissolved in 10 μl of 1,3-butanediol, and 300 μl of the incubation medium is added. This mixture is placed in the thermostated waterbath, and the incubation is started by the addition of mitochondria or mitoplasts at a final protein concentration of 2.2 mg/ml. After 10 min, the reaction is stopped by adding 150 μl of concentrated sample buffer followed by an incubation at 95°C for 3 min. SDS-polyacrylamide gel electrophoresis (SDS-PAGE) is performed on slab gels according to Laemmli (1970) as modified (Douglas and Butow 1976) using a 10% polyacrylamide gel (w/v). Up to 150 μg of protein may be applied per lane. After electrophoresis, the gels are either stained for protein, or impregnated with salicylate, and processed for autoradiography (Chamberlain 1979), and exposed for several days to X-ray film. Protein was visualized using Coomassie blue (0.25% Coomassie blue (w/v) - 7.5% acetic acid (v/v), and 50% methanol in water (v/v)) for staining, and 7.5% acetic acid (v/v) - 20% methanol (v/v) for the destaining of the background. Figure 1 illustrates the result of an experiment which was performed to compare crude mitochondria with highly purified mitoplasts in respect to the pattern of protein (P) and radioactivity (A) after incubation with labeled myristate. Mitochondria and highly purified mitoplasts do not significantly differ in their acylation pattern showing that the acylation occurs, indeed, in isolated mitochondria, and is not due to contamination by other organelles.

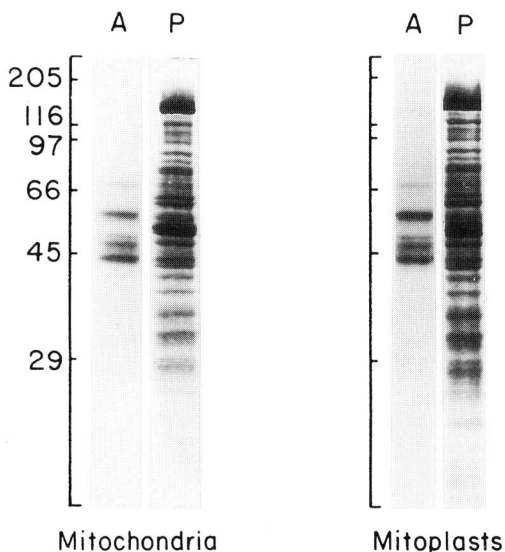

Fig. 1. Acylation of isolated mitochondria. Mitochondria were isolated and incubated with 75 μM [^{14}C]myristic acid for 10 min. Radioactivity incorporated into mitochondrial proteins was visualized by subjecting the denatured proteins to SDS-PAGE and subsequent autoradiography. Parallel lanes were stained with Coomassie blue (P), or processed for autoradiography, and exposed to X-ray film for 4 days (A). On the left (control), mitochondria isolated by the normal procedure, were used. On the right, the result obtained with mitoplasts is shown. The scale indicates relative molecular weights (M$_r$ x 10^{-3}).

Experiment B:

Chain length specificity of mitochondrial acylation and deacylation

The fatty acid chain length specificity of the acylation reaction may be demonstrated by incubation of the mitochondria for 5 min with 80 μM unlabeled fatty acid with differing chain length in different incubations prior to acylation with 25 μM radioactive myristate. Acylation with unlabeled acids will be detected after autoradiography as a decrease in radioactive labeling with radioactive myristate as compared to a control reaction in the absence of cold acid.

Chain length specificity of the deacylation may be shown in a similar experiment. Mitochondria are acylated using 25 μM radioactive myristate, and after 5 min, unlabeled acid (80 μM) is added, and the reaction stopped by heat inactivation after another incubation period of 5 min.

Experiment C:

Identification of the radioactivity incorporated into proteins

Lanes from fresh gels are incubated in either 1 M Tris, pH 9.9 (control), 1 M hydroxyla-mine, pH 9.9, or 1 M KOH in 20% methanol in water (v/v), and hydrolyzed at room tempera-ture for 2 h. Subsequently, the lanes were processed as above in order to visualize by auto-radiography the remaining radioactivity after the different conditions for hydrolysis.

The chemical identity of the radioactivity may be determined following a similar proce-dure (Schlesinger et al. 1980). Briefly, slices corresponding to labeled bands are cut from fresh unfixed gel, homogenized, and incubated in 2 ml hydroxylamine for 24 h. This hydroly-sis is carried out on a shaker at room temperature. The resulting suspension is centrifuged at 10,400 g_{max}, and the supernatant containing the hydrolysate is extracted three times with 2 ml chloroform - methanol (2:1 v/v). The pooled organic solvent extract is evaporated under a stream of nitrogen gas, and the residue can be analyzed on thin layer chromatography plates according to Heusser (1968) using methanol - dioxan - chloroform - 50 mM glycine, pH 3.0, in water (4:3:2:3) as the mobile phase. Radioactive fatty acid(s) may conveniently be used as standards. The dried thin-layer plates were impregnated with En3hance spray and exposed to X-ray film.

COMMENTS

As in the case of acylation occurring in the endoplasmic reticulum, the role of mito-chondrial acylation is far from clear. We have suggested a possible involvment in the regula-tion of oxidative phosphorylation (Stucki et al. 1988, submitted for publication) and believe on the basis of further experiments to be able to exclude artifacts due to possible interme-diates of fatty acid oxidation (Stucki et al. 1988, submitted for publication). Whatever the role, the presence in mitochondria of covalent protein modification by fatty acids exhibiting broad chainlength specificity should be borne in mind when discussing patterns obtained from whole cell acylation experiments.

Acknowledgment. This work was supported by grants of the Swiss National Science Founda-tion.

REFERENCES

Aebi H (1970) In: Bergmeyer HU (ed) Methoden der enzymatischen Analyse. Verlag Chemie Weinheim, pp 636-641

Baginski ES, Foa PP, Zak B (1970) In: Bergmeyer HU (ed) Methoden der enzymatischen Analyse. Verlag Chemie Weinheim, pp 839-841

Batayneh N, Kopacz SJ, Lee CP (1986) Arch Biochem Biophys 250:476-487

Baum H, Dogson KS, Spencer B (1959) Clin Chim Acta 4:453-455

Chamberlain JP (1979) Analyt Biochem 98:132-135

Douglas M, Butow RA (1976) Proc Natl Acad Sci USA 73:1083-1086

Heusser D (1968) J Chromatogr 33:62-69

Johnson D, Lardy HA (1967) In: Estabrook RW (ed) Methods in Enzymology, Vol 10. Academic Press New York London, pp 94-96

Laemmli UK (1970) Nature 227:680-685

Peterson PL, Greenawalt JW, Reynafarje B, Hullien J, Decker GL, Soper JW, Bustamante E (1978) Meth Cell Biol 20:411-481

Schlesinger MJ, Magee IA, Schmidt MGF (1980) J Biol Chem 255:10021-10024

Schmidt E (1970) In: Bergmeyer HU (ed) Methoden der enzymatischen Analyse. Verlag Chemie Weinheim, pp 607-613

Schnaitmann C, Erwin VG, Greenawalt JW (1961) J Cell Biol 32:719-735

Sefton BM, Buss JE (1987) J Cell Biol 104:1449-1453

Stucki JW, Lehmann L, Sigel E (1988) submitted for publication

Wojtczak L (1976) J Bioenerg Biomembr 8:293-311

Plant Protein Acylation: Identification of the Modified Proteins and Analysis of the Bound Fatty Acid Ligand

A.K. Mattoo*, H.A. Norman[+], F.E. Callahan*, J.B. St.John[+], and M. Edelman*[#]

*Plant Hormone, and [+]Weed Science Laboratories, USDA/ARS,
Beltsville Agricultural Research Center (W), Beltsville, Md. 20705, USA
[#]Permanent address: Plant Genetics Department, The Weizmann Institute of Science,
IL-76100 Rehovot, Israel

INTRODUCTION

Post-translational covalent attachment of fatty acids to several proteins was recently demonstrated in intact plants of *Spirodela oligorrhiza*, an aquatic angiosperm (Mattoo and Edelman 1987). Protein acylation occurred within 1-3 min of radiolabeling, was lightly stimulated, and was localized to the photosynthetic organelle (the chloroplast) (Mattoo and Edelman 1987; Mattoo et al. submitted for publication). The nuclear coded 10 kDa acyl carrier protein and 26 kDa light harvesting chlorophyll a/b apoprotein, and the chloroplast coded large subunit of ribulose-1,5-bisphosphate carboxylase/oxygenase and D2, and 32 kDa photosystem II reaction center proteins were prominent among the [^{3}H]palmitate labeled chloroplast proteins. Only the acyl-linkage with acyl carrier protein was determined to be linked via a thiol ester bond; the remaining acylated proteins appeared to be linked via ether or amide bonds (Mattoo and Edelman 1987).

In this chapter, we present methods for *in vivo* acylation of plant proteins and identifying the radiolabeled moiety in the protein bands on nitrocellulose paper after SDS-polyacrylamide gel electrophoresis.

EXPERIMENTAL PROCEDURES

In vivo labeling of proteins with [^{3}H]fatty acids

Axenic *Spirodela oligorrhiza* (Kurtz) Hegelm plants were grown phototrophically at 25^0-28^0C under cool white fluorescent lamps (20 μmol.m^{-2}.s^{-1}) for 10-15 days on half-strength

Hutner's mineral medium (Posner 1967) containing 0.5% sucrose. For pulse labeling, the plants were transferred for 24-48 h to 35 x 10 mm sterile petri dishes (Falcon style) containing mineral medium lacking sucrose. Radiolabeling was initiated with 600-1000 μCi of [9,10-^{3}H(N)]palmitic acid (> 20 Ci/mmol) or [9,10-^{3}H(N)]myristic acid (> 30 Ci/mmol) (New England Nuclear) per ml of mineral medium. Prior to use, the radioactive fatty acid solutions were dried under a stream of nitrogen gas and then dissolved in dimethyl sulfoxide. The concentration of dimethyl sulfoxide in the labeling medium was kept under 1%. After labeling for 1-3 min, the plants were washed with 10 ml of mineral medium and immediately frozen on dry ice.

Cell homogenization and membrane fractionation

The methods used for the isolation of soluble and whole cell membrane proteins (Reisfeld et al. 1982), and chloroplast membranes (Mattoo et al. 1981) were the same as previously described.

SDS-PAGE of acylated proteins

Samples for gel electrophoresis were prepared by adding 0.5 volumes of application buffer (Mattoo et al. 1981) to soluble fractions, and 1.0 volume of application buffer to membrane fractions. Soluble samples were heated at 90°C for 5 min, while membrane samples were incubated at 25°C for 1-2 h, and centrifuged at 12,000 x g for 30 sec prior to application on gel slots. Electrophoresis was carried out according to Laemmli (1970) using a 5% stacking gel and a 10-20% gradient resolving gel (Marder et al. 1986). Following electrophoresis, the gels were stained and fixed with Coomassie blue R-250 in 50% methanol - 7% acetic acid for 60 min, destained overnight in 20% methanol - 7% acetic acid, fluorographed using En^{3}hance (New England Nuclear), dried and exposed on Kodak X-AR5 X-ray films (Marder et al. 1986).
The SDS-polyacrylamide gel pattern of rapidly acylated soluble and membrane plant proteins is shown in Figure 1.

Electroblotting of radiolabeled proteins onto nitrocellulose paper

Transfer of electrophoretically resolved proteins in SDS-polyacrylamide gels onto nitrocellulose paper (0.1-0.2 μm pore size, Schleicher and Schull) was carried out essentially as described (Towbin et al. 1979) using 25 mM Tris, 192 mM glycine, 20% methanol, 0.02% SDS, as the transfer buffer. Electroblotting was carried out at 40 V overnight, the temperature of the transfer buffer being maintained at 10°-20^oC. After transfer, the nitrocellulose paper was washed in phosphate-buffered saline for 5 min, air dried, and exposed to Kodak X-AR5 X-ray film to locate the radioactive bands. [^{35}S]Methionine labeled *Spirodela* extracts, and [^{14}C]protein standard markers (Amersham) were run alongside [^{3}H] labeled samples to locate the protein bands on the nitrocellulose paper.

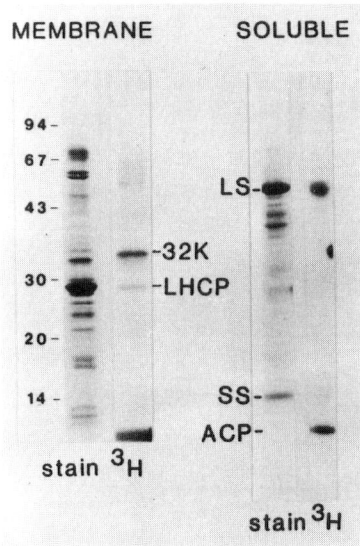

Fig. 1. Rapidly acylated soluble and membrane proteins of *Spirodela oligorrhiza*. Plants were labeled *in vivo* for 3 min with [^{3}H]palmitic acid. Soluble and membrane proteins were isolated and fractionated by SDS-PAGE as described in the text. Steady-state protein pattern (stain) and pulse labeled protein pattern (^{3}H) are shown. The positions of the large subunit (LS), and the small subunit (SS) of ribulose-1, 5-bisphosphate carboxylase/oxygenase, the 10 kDa acyl carrier protein (ACP), the light harvesting chlorophyll a/b apoprotein (LHCP), and the 32 kDa photosystem II reaction center protein (32K) are indicated.

Identification of the labeled moiety on acylated plant proteins

1. Extraction

Nitrocellulose strips containing the resolved radiolabeled protein band of interest, were extracted with 5 ml of chloroform - methanol (2:1) to remove any non-covalently bound lipid. The strips were then cut into smaller pieces and refluxed at 70^oC for 8 h with 3 ml of

methanol - 3 N sodium hydroxide (9:1). After cooling to 25°C, the reaction solution was acidified to pH 1.0-2.0 with 6 N HCl, and extracted three times with 5 ml of petroleum ether. The
organic phases were combined and washed three times with equal volumes of water. After adding 200 μl of ethanol, the solution was dried completely under nitrogen. The preparations
were then immediately redissolved in 0.5 ml of hexane and transferred to 1 ml glass 'Reacti-
Vials' with teflon-lined screw caps (Pierce Chem. Co., Rockford, IL). Aliquots of the hexane
solution, and the washes were counted for radioactivity in a liquid scintillation spectrome-
ter. No radioactivity was detected in the washes.

2. Separation by HPLC

The fatty acids recovered were converted to p-nitrophenacyl derivatives for resolution by
HPLC (Jordi 1978; Halgunset et al. 1982). The derivatizing reagents were p-nitrophenacyl
bromide (Aldrich Chem. Co., Milwaukee, WI), diisopropyl ethylamine (Sigma Chem. Co., St.
Louis, MO), and N,N-dimethylformamide (Fisher Scientific). Diisopropyl ethylamine was
dried over sodium hydroxide prior to use, and N,N-dimethylformamide was stored with oven-
activated molecular sieves from which the dust had been removed with nitrogen. A solution
of 20 μmol p-nitrophenacyl bromide and 40 μmol diisopropyl ethylamine in N,N-dimethylfor-
mamide was used.

Hexane solution containing fatty acids in glass vials, was evaporated under nitrogen.
80 μl of the combined derivatizing reagents were added to the residue. Vials were then heated at 65°C for 15 min. Derivatization was 92-95% as assessed by thin-layer chromatography
of the fatty acid derivatives on silica gel H using petroleum ether - diethyl ether - acetic
acid (70:30:1) as the solvent system. Fatty acids (R_f 0.35) and p-nitrophenacyl derivatives (R_f
0.21) resolved by this system, were identified by exposure of the thin-layer plate to iodine
vapor. Lipid bands were marked, and after allowing the iodine vapor to dissipate, transferred to scintillation vials suspended in 10 ml Aquasol-2 (Du Pont), and counted for radioactivity.

Separation of the p-nitrophenacyl fatty acids was achieved using an HPLC system (Waters
Assoc., Model 6000A) equipped with a model U6K universal injector, and a 15 cm x 4.5 mm
Ultrasphere-ODS (5 μm) reverse-phase column (Alltech Assoc. Inc., Deerfield, IL). 50 μl con-
sisting of 20 μl derivatized sample, and 30 μl mobile phase were injected. The mobile phase
was methanol - acetonitrile - water (82:9:9) with a flow rate of 1 ml/min (Jordi 1978). The
samples were eluted first through a detector monitoring at 254 nm (Waters Assoc., Model 450
Variable Wavelength Detector), and then eluted through a Flow-One Beta (model IC) HPLC
Radioactive Flow Detector (Radiomatic Instruments and Chemical Co. Inc., Tampa, FL)
using Flow-Scint III (Radiomatic Instruments and Chemical Co. Inc.) (as scintillator), mixed
with the HPLC eluate by means of a pump adjusted to give a flow rate of 2 ml/min (Norman
and St.John 1986). Peak identity was determined by co-elution with authentic derivatives of

fatty acid standards. Also, 0.5 ml fractions eluting every 0.5 min were collected, dried, and counted in Toluene-Liquifluor (Du Pont).

In Figure 2, the identity of the ^{3}H labeled material extracted from the 32 kDa photosystem II reaction center protein is shown to be palmitic acid. Inability to detect any ^{3}H]myristic acid on the 32 kDa protein when plants were labeled with ^{3}H]myristic acid, demonstrates specific palmitoylation of this protein.

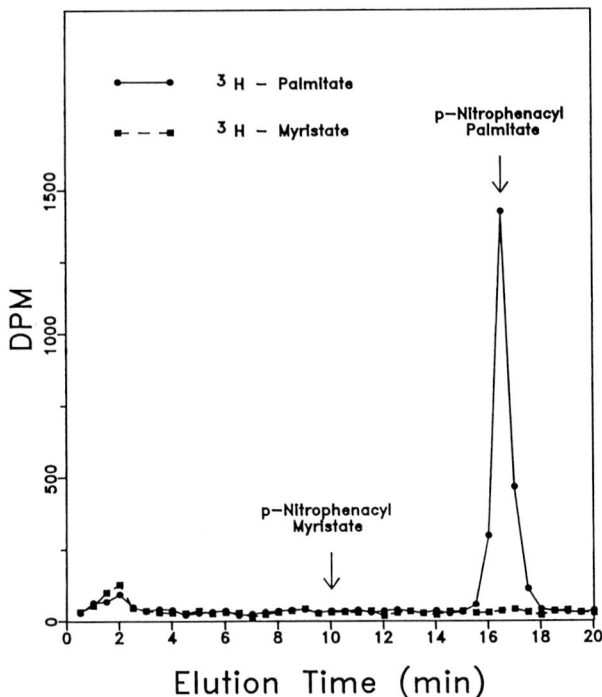

Fig. 2. Reverse-phase HPLC separation of [3H]p-nitrophenacyl derivatives of fatty acids released from the 32 kDa protein. Plant proteins acylated *in vivo* for 2 min with [3H]palmitate or [3H]myristate as described in the text, were fractionated by SDS-PAGE, and the gel blotted onto nitrocellulose paper. The region of the 32 kDa protein was located on the blot and cut out. Fatty acids were then extracted, derivatized, and fractionated by HPLC. Eluants were collected at 0.5 min intervals and counted for radioactivity. The distribution of radioactivity was compared with the elution of p-nitrophenacyl fatty acid standards whose positions are marked by arrows.

Acknowledgments. We thank Cathy Conlon and Roshni Mehta for excellent technical assistance. This investigation was supported in part by a United States - Israel Binational Agricultural Research and Development (BARD) grant to A.K. Mattoo and M. Edelman. Mention of specific instruments, trade names, or manufacturers is for the purpose of identification and does not imply an endorsement by the United States Government.

REFERENCES

Halgunset J, Lund EW, Sunde A (1982) J Chromatogr 237:496-499
Jordi HC (1978) J Liq Chromatogr 1:215-230
Laemmli UK (1970) Nature 227:680-685
Marder JB, Mattoo AK, Edelman M (1986) In: Weissbach A, Weissbach H (eds) Methods in Enzymology, Vol 118, Academic Press New York London, pp 384-396
Mattoo AK, Pick U, Hoffman-Falk H, Edelman M (1981) Proc Natl Acad Sci USA 78:1572-1576
Mattoo AK, Edelman M (1987) Proc Natl Acad Sci USA 84:1497-1501
Mattoo AK, Callahan FE, Mehta RA, Ohlrogge JB (submitted for publication)
Norman HA, St.John JB (1986) Plant Physiol 81:731-736
Posner HB (1967) In: Witt FA, Wessels NK (eds) Methods in Developmental Biology. Crowell New York, pp 301-317
Reisfeld A, Mattoo AK, Edelman M (1982) Eur J Biochem 124:125-129
Towbin H, Staehelin T, Gordon J (1979) Proc Natl Acad Sci USA 76:4350-4354

Identification and Characterization of Lipid-Modified Membrane Proteins in Bacteria

S. Hayashi and H.C. Wu

Department of Microbiology, Uniformed Services University of the Health Sciences, Bethesda, Maryland 20814-4799, USA

INTRODUCTION

The existence of lipid-modified protein in bacteria was first demonstrated by the discovery of murein lipoprotein in *E. coli* (Braun and Rehn 1969). Like other exported proteins in *E. coli*, the outer membrane lipoprotein is first synthesized as a precursor, the prolipoprotein, and subsequently processed to form mature lipoprotein (Inouye et al. 1977). A novel cyclic peptide antibiotic, globomycin, specifically inhibits the processing of prolipoprotein, and causes the accumulation of glyceride-modified prolipoprotein (Hussain et al. 1980). The use of globomycin as a specific inhibitor of prolipoprotein processing in conjunction with metabolic labeling of cell envelope proteins with radioactive lipid precursors ($[^3H]$palmitate or $[2-^3H]$glycerol) has led to the identification of many lipid-modified membrane proteins in bacteria. They all contain an N-acyl-diglyceride-cysteine residue at their NH_2-termini as was first found in the murein lipoprotein (Hantke and Braun 1973). The precursor forms of these lipid-modified membrane proteins contain a consensus tetrapeptide sequence Leu-X-Y-Cys at the modification and processing site located at the COOH-terminal region of the signal sequence. Experimental procedures used for the identification and characterization of lipid-modified membrane proteins in bacteria are described.

EXPERIMENTAL PROCEDURES

Experiment 1
Labeling of lipid-modified proteins with $[2-^3H]$glycerol or $[^3H]$palmitate

1. Labeling of lipoprotein

Medium: Enriched broth medium such as proteose peptone beef extract (PPBE) broth medium (Wu and Wu 1971) or L broth medium (Miller 1972).

Bacterial cells are inoculated into 5 ml of PPBE medium and incubated at 37°C in a gyrotory waterbath (200-300 rpm) overnight. The overnight preculture is used to inoculate 25 ml of PPBE medium in a 125 ml Erlenmeyer flask. Adjust the initial A_{600nm} to 0.1- 0.15. Incubate the culture in a gyrotory waterbath at 37°C. Check the A_{600nm} periodically, and at A_{600nm} 0.2-0.25, 100 μCi of either [2-^{3}H]glycerol (sp. act. 5-10 Ci/mmol), or [^{3}H]palmitate (sp. act. 10-30 Ci/mmol), is added to the culture. When the A_{600nm} of the culture reaches 0.9-1.0, the cells are harvested in a refrigerated centrifuge (4°C) at 5000 x g for 10 min.

2. Preparation of the cell envelope

The harvested cells are resuspended in 5 ml of 10 mM phosphate buffer, pH 7.0, sonicated in a Sonifier cell disruptor at 0°C, and the crude extract is centrifuged at 1500 x g at 4°C for 10 min in order to remove the unbroken cells. The supernatant fraction is transferred to a screw-capped tube, centrifuged at 250,000 x g at 4°C for 2 h, and the supernatant solution is carefully removed with a Pasteur pipette. The pellet which contains the cell envelope fraction, is resuspended in 1 ml of water, transferred into a glass conical centrifuge tube, and lyophilized. 2 ml of chloroform - methanol (2:1) is added to the lyophilized cell envelope, vortexed vigorously, and centrifuged in a table-top centrifuge at 2000 x g for 15 min. The chloroform/methanol extract is removed carefully with a Pasteur pipette and the delipidation procedure is repeated at least three times. The cell envelope residue is dried in a heating block, and the dried cell envelope is solubilized in 500 μl of 10 mM phosphate buffer, pH 7.0, containing 1% SDS, with the aid of a bath sonicator, and incubated at 100°C for 5 min for the solubilization of membrane proteins.

3. Immunoprecipitation and SDS-PAGE analysis

100-200 μl of the solubilized membrane fraction is placed in an 1.5 ml Eppendorf tube, and 900 μl of 10 mM phosphate buffer, pH 7.0, is added. Immunoprecipitation of the membrane lipoprotein is carried out as follows:

70 μl of *Staphylococcus* cell suspension (Staph A, Calbiochem, CA) is added to 1 ml of the diluted membrane solution, incubated at 4°C for 10-15 min, and the solution is centrifuged in a microfuge. The supernatant solution is carefully removed with a Pasteur pipette and transferred to a clean 1.5 ml Eppendorf tube. 50 μl of anti-lipoprotein serum is added. Following an incubation at 4°C for 20 min, 50 μl of Staph A suspension is added, and the mixture is further incubated at 4°C for 20 min, and then centrifuged in a microfuge. The Staph A pellet is washed at least twice with NET buffer (Kessler 1975), and the washed Staph A pellet is suspended in 15 μl of SDS-polyacrylamide gel electrophoresis sample buffer (Inouye and Guthrie 1964). The suspension is heated at 100°C in a heating block (or boiling waterbath) for 3-5 min, centrifuged in a microfuge, and the supernatant solution is applied for

SDS-PAGE (Inouye and Guthrie 1964). After electrophoresis, the gel is fixed and stained with Coomassie blue. Fluorography is then carried out by soaking the slab gel in Amplify (Amersham) or En^{3}hance (New England Nuclear) for 20-30 min with gentle shaking. The treated slab gel is dried in a gel dryer and exposed to X-ray film.

Experiment 2

Accumulation of prolipoprotein in globomycin-treated cells

As mentioned in the Introduction, the signal peptidase for prolipoproteins is specifically inhibited by globomycin. This finding has been exploited to ascertain whether a particular protein is lipid-modified or not. In this experiment, the method used for identification of *E. coli* outer membrane prolipoprotein is described.

Bacterial cells are inoculated into 5 ml of M9-glucose minimal medium (Miller 1972), and incubated in a waterbath shaker at 37°C overnight. Inoculate the overnight culture into 5 ml of M9-glucose medium at an initial A_{600nm} of 0.2-0.3, and incubate the culture in a water-bath shaker at 37°C. At the middle log phase (A_{600nm} of 0.5-0.6), 1 ml aliquots of the culture are placed into 1.5 ml Eppendorf tubes. Globomycin* (10 mg/ml in DMSO) is added to a final concentration of 0 (control), 50 μg/ml, and 100 μg/ml, respectively. The cultures are mixed well, incubated at 37°C for 5 min, and labeled with 50 μCi of [^{35}S]methionine (sp. act. 100 mCi/mmol) for 5 min. Labeling is terminated by adding 100 μl of 50% trichloroacetic acid (TCA). Following a 15 min incubation on ice, the TCA precipitates are collected in a microfuge, washed with acetone once, and dried in a heating block. The dried pellets are re-suspended in 200 μl of 10 mM phosphate buffer, pH 7.0, in a bath sonicator, and incubated at room temperature for 15 min after successively adding 5 μl of 1 N NaOH and 60 μg of lysozyme. SDS (final concentration 1%) is added into each tube, and the mixture is heated in a heating block for 5 min. 100 μl of this solution is used for immunoprecipitation, and the immunoprecipitate is analyzed by SDS-PAGE as described in Experment 1, section 3.

Bacterial lipoproteins are characterized by the presence of N-acyl-diglyceride-cysteine residue at their NH$_2$-termini as mentioned in the Introduction. Experiments 3 and 4 describe the identification of glyceryl-cysteine residue in the lipoprotein, and the quantitation of amide- and ester-linked fatty acids, respectively. The method for the purification of radio-isotopically labeled lipoprotein is described first.

---

* Globomycin is not available as a commercial product. It is a gift from Dr. M. Arai, Sankyo Pharmaceutical Co., Ltd., Tokyo, Japan.

Experiment 3

Identification of glyceryl-cysteine residue in lipoprotein

1. Purification of labeled protein

M9-glucose medium supplemented with methionine (200 μg/ml) and cysteine (10 μg/ml) is used for the steady-state labeling of bacterial cells with 200-300 μCi [^{35}S]cysteine (sp. act. 250 mCi/mmol). 500 μl of SDS solubilized cell envelope is immunoprecipitated and analyzed by SDS-PAGE. After autoradiography, the lipoprotein band is sliced from the dried gel, and soaked in small volume (3-5 ml) of Laemmli gel running buffer (Laemmli 1970). The gel pieces automatically come off from the filter paper. Pour the gel pieces and the buffer into a dialysis tube (Spectrapor, M.W. cutoff 3500). Place the dialysis tube in a horizontal mini-submarine gel apparatus filled with the Laemmli gel running buffer. The labeled protein is eluted from the gel pieces by electrophoresis at a constant voltage of 150-200 V for 2 h, and this procedure is repeated twice. Protein elution is monitored by the elution of Coomassie blue dye from the gel pieces. The solution containing the labeled protein and the blue dye is dialyzed against water for at least three days with three changings of water in order to remove the excess salt, glycine, and SDS. The dialyzed solution is lyophilized.

2. Identification of glyceryl-cysteine in lipoprotein

Purified [^{35}S]cysteine labeled lipoprotein is dissolved in 1 ml of performic acid, and kept at 4°C overnight. Following performic acid oxidation, the sample is lyophilized to remove performic acid. 1 ml of 6 N constant boiling hydrochloric acid is added into the hydrolysis tube, evacuated, and sealed in vacuo. The sample is kept at 110°C for 20 h. The hydrolysate is transferred into a glass conical centrifuge tube, lyophilized to remove hydrochloric acid, and the dried sample is analyzed by high-voltage paper electrophoresis.

The lyophilized sample is dissolved in a small volume (usually 20-30 μl) of electrophoresis running buffer (0.47 M formic acid, 1.4 M acetic acid, pH 2.3-2.4), and 5-10 μl of this sample is applied onto the center of a 60 cm Whatman 3MM paper. Cysteic acid, cysteine, methionine sulfone, and methionine are applied as standards. Following electrophoresis at 100 V/cm for 2-3 h, the paper is dried, and sprayed with ninhydrin solution to identify the position of the standard compounds. The sample lane is sliced at 1 cm intervals, and the radioactivity in each slice of paper is determined by liquid scintillation counting. Glyceryl-cysteine migrates at 5-10 cm towards cathode, while cysteic acid derived from the cysteine residue of protein migrates at 10-15 cm towards anode (Lin and Wu 1976).

Experiment 4

Identification of amide- and ester-linked fatty acids in lipoprotein

Labeling of lipoprotein with [^{3}H]palmitate is carried out as described in Experiment 1. Purification of [^{3}H]palmitate labeled lipoprotein is carried out as described for [^{35}S]cysteine labeled lipoprotein in Experiment 3. The gel is stained with Coomassie blue to locate the lipoprotein band. To preserve the labile ester-linked fatty acids in lipoprotein, the gel is sliced immediately after destaining without drying the gel under acidic conditions.

Purified [^{3}H]palmitate labeled protein is resuspended in 100 μl of water, and the total radioactivity in the sample is determined by counting 5 μl of the suspension. The aqueous suspension is divided into two equal parts and lyophilized. One sample is dissolved in 15 μl of SDS-PAGE sample buffer, and designated the control without alkali treatment (-OH). 8 μl of 0.1 N NaOH solution is added to the other sample, and the mixture is incubated in a waterbath at 37°C for 2 h with occasional mixing. Following the 2 h incubation, 2 μl of 0.4 N HCl solution is added followed by adding 5 μl SDS-PAGE buffer. This sample is designated alkali-treated (+OH). Both -OH and +OH treated samples are incubated in a heating block at 95°C for 5 min, and 2 μg of cytochrome C is added to each sample as a molecular weight marker. The outer membrane lipoprotein (M.W. 7500) migrates between the dye front and cytochrome C (visible without staining) in SDS-PAGE. After electrophoresis, the gel is dried, and the region of the gel between cytochrome C and 5 mm below the dye front is sliced at 1 mm intervals (about 25-30 sections). Each gel piece is placed in a plastic vial, and incubated in 0.5 ml of 1% SDS solution at 37°C overnight. The radioactivity of each fraction is determined by liquid scintillation counting. The +OH treated sample will give a peak of free [^{3}H]palmitate (released by alkali treatment) at the dye front, and a second peak between cytochrome C and dye. The palmitate released by +OH treatment represents ester-linked fatty acids of lipoprotein, and the radioactivity in the second peak (lipoprotein) represents the amide-linked fatty acid of the lipoprotein. In -OH treated control lane, only one [^{3}H]palmitate labeled peak migrates at the same position as the second peak (lipoprotein) of +OH treated sample. Integration of radioactive counts of both peaks in the +OH treated sample will reveal that the percentage of the alkali-released palmitate is around 40% (the sum of radioactivities of both peaks in +OH treated sample should be equal to that of single peak in -OH treated sample).

Acknowledgments. This work was supported by grants from the National Institutes of Health (GM-28811), and the American Heart Association (84-606).

REFERENCES

Braun V, Rehn K (1969) Eur J Biochem 10:426-438
Hantke K, Braun V (1973) Eur J Biochem 34:284-296
Hussain M, Ichihara S, Mizushima S (1980) J Biol Chem 255:3707-3712
Inouye M, Guthrie JP (1964) Proc Natl Acad Sci USA 64:957-961
Inouye S, Wang S, Sekizawa J, Halegoua S, Inouye M (1977) Proc Natl Acad Sci USA 74: 1004-1008
Kessler SWC (1975) J Immunol 115:1617-1624
Laemmli UK (1970) Nature 227:680-685
Lin JJC, Wu HC (1976) J Bacteriol 125:892-904
Miller JH (1972) Experiments in Molecular Genetics, Cold Spring Harbor Laboratory, pp 431
Wu HC, Wu TC (1971) J Bacteriol 105:455-466

Detection of Palmitoylating Activity with Exogenous Viral Acceptor Proteins

M.F.G. Schmidt, G.R. Burns*, M. Berger[+], and M. Schmidt[#]

*Department of Biochemistry, Faculty of Medicine, Kuwait University, P.O. Box 24923
13110 Safat, Kuwait, Arabian Gulf
[+]Abteilung D/NPB2, Boehringer AG, Bahnhofstrasse 9-15, D-8132 Tutzing,
Federal Republic of Germany
[#]Institut für Virologie, Justus-Liebig-Universität, Frankfurter Strasse 107,
D-6300 Giessen, Federal Republic of Germany

INTRODUCTION

Fatty acylation has become a modification of proteins which seems now almost as widespread in nature as glycosylation (for reviews see Schmidt 1983; Sefton and Buss 1987). Two types of acylation are distinguished: palmitoylation of proteins into internal ester-linkages to serine, threonine, or cysteine, and myristoylation which usually occurs in amide-linkage on amino terminal glycin residues (Sefton and Buss 1987). Whereas palmitoylation has been reported to be reversible, and thus, can be of regulatory value for a cell's metabolism, the latter modification occurs only once in the life cycle of a given myristoylated acyl protein (Magee and Courtneidge 1985; McIlhinney et al. 1985; Berger and Schmidt 1986). An enzyme for the myristoylation of proteins and peptides has been characterized recently from yeast cells (Towler et al. 1987). The isolation of enzymes for palmitoylation has not yet been achieved, although such an activity has been reported to occur in various biological materials (kidney, liver, cultured cells of various origin). Therefore, it seems useful to describe the procedure which allows to measure palmitoylating activity.

EXPERIMENTAL PROCEDURES

For the palmitoylation of proteins, the virus system has been found to be useful since cell culture and all facilities necessary are available to grow and purify viruses with acylated proteins to large quantities. Cell culture also has the advantage over tissue samples for the preparation of microsomal membranes.

Purification of virus

Relevant enveloped viruses (e.g., Semliki Forest virus, SFV; vesicular stomatitis virus, VSV) are grown in cultured baby hamster kidney cells (BHK) until a cytopathic effect becomes apparent. Medium with virus particles is centrifuged at 2000 g at 4°C to remove cell debris. Centrifugation at 19,000 rpm in the large volume Beckman 19 rotor for 3 h will suffice to yield virus sediments necessary for both structural studies of the acylation site (in this case, the virus particles would be labeled with [^{3}H]palmitic acid according to Schmidt 1982) and the use as acceptor during fatty acylation *in vitro*.

Preparation of microsomes

Confluent cultures of BHK or chicken embryo fibroblasts (usually 30 dishes with a diameter of 15 cm) are washed and scraped with ice-cold EDTA-buffer (0.14 M NaCl, 70 mM KCl, 1 mM KH_2PO_4, 10 mM EDTA, pH 7.4). They are then centrifuged at 1000 g, pellets washed with ice-cold Tris-buffer (20 mM Tris-HCl, 0.15 M NaCl, pH 7.4), and centrifuged again. The sediments are taken up in a small volume of Tris-buffer, and the suspension is added in drops to 9 volumes of a hypotonic buffer solution (20 mM Tris-HCl, pH 7.4) in an all glass homogenizer and incubated at 4°C for 20 min. The cells are then homogenized by 20 strokes (tight-fitting homogenizer). Thereafter, 0.1 volumes of a solution of 30 mM $MgCl_2$ - 100 mM NaCl are added, and the mixture homogenized by another 3 strokes. Nuclei and unbroken cells are sedimented at 1000 g. The supernatant is centrifuged at 100,000 g, usually in a Beckman 60 Ti-rotor at 4°C for 1 h, and the resulting pellet suspended in a small volume of phosphate buffered saline (PBS), pH 7.2, or in TNA-buffer (20 mM Tris-HCl, pH 7.4, 0.15 M NaCl, 10 mM EDTA) to obtain a protein concentration of 10-15 mg/ml. At any step of the procedure, the material has to be kept at 4°C.

Deacylation of viral acceptor proteins

Since the viruses contain acylated proteins, fatty acids have to be released before the preparation can be used as acceptor for radiolabeled acyl chains to be incorporated into the respective proteins *in vitro*. Virus particles (about 10 mg protein) are disrupted by treatment with 1% Nonidet P40 at room temperature for 10 min. Fatty acid release from viral proteins E1, E2 (of SFV), and G-protein (of VSV) is achieved by adding hydroxylamine, pH 6.5, to solubilized virus to give a final concentration of at least 0.5 M NH_2OH and stirring at room temperature for 4 h. If virus particles with different acyl proteins are used, it is advisable to verify de facto deacylation by using [^{3}H]palmitic acid labeled preparations.

Practical Procedure

1. Reaction mixture for palmitoylation *in vitro*

The following components are mixed in Eppendorf vials, or better in 10 ml glass conicals, all kept in an ice-bath.

N⁰	Microsomes	Deacylated Acceptor Protein	Pretreatment/ Additions	[^{14}C]Pal- Coenzyme A[d]	TNE- buffer
1	0 µl	30 µl	–	20 µl	200 µl
2	2	30	–	20	198
3	5	30	–	20	195
4	10	30	–	20	190
5	30	30	–	20	170
6	10	30	preheat micro- somes at 60⁰ for 10 min[c]	20 µl	190
7	10	30	preheat acceptor at 60⁰ for 10 min[c]	20	190
8	10	30	add 5 µl SDS [10%]	20	185
9	10	30 µl solubilized virus (not deacy- lated)[a]	–	20	190
10	10	30 µl mock- acceptor[b]	–	20	190

a) Samples of virus disrupted with 1% NP40.
b) 1% NP40, 0.5 M NH_2OH in TNE.
c) Pretreatment of microsomes/acceptor protein after suspending in TNE.
d) [1-^{14}C]palmitoyl Coenzyme A (0.4 µCi) from NEN is added last.

The reaction is started by adding the acyl donor (in this case [^{14}C]palmitoyl CoA), and incubated at 28°C for 30 min. The reaction is terminated by adding at least 1 ml of chloroform - methanol (1:2; v/v) which precipitates proteins, and at the same time extracts the labeled substrate and its lipid products as well as hydroxylamine (originating from the acceptor preparation) present in the mixture. Proteins are sedimented at 3000 rpm for 5 min, resuspended in 1 ml of methanol, and centrifuged again. Supernatants are retained for TLC-analysis on silica plates using chloroform - methanol - water (60:35:8; v/v) as a solvent, and protein sediments are solubilized in sample buffer (20% glycerol, 4% SDS, 125 mM Tris-HCl, pH 6.8, 0.004% bromphenol blue) for polyacrylamide gel analysis (PAGE) on 10% SDS gels. Authentic virus samples are run as reference in parallel lanes.

2. Detection of radioactivity

Since levels of incorporation are quite low, autoradiography (fluorography) of gels takes at least 2-3 weeks. In order to avoid such long exposure times, it has been found useful to measure radioactivity directly in the relevant gel regions. After Coomassie staining, the gel bands of E1, E2 (SFV), and G-protein (VSV) (or other potential acceptor proteins) are excised and cut into pieces of about 3 x 3 mm^2, transferred into counting vials, and frozen at -80°C for 30 min. To solubilize the protein, 500 μl of Soluene 350 (Packard) is added, and the samples stirred. Tubes are sealed and incubated at 60°C for 2 h, or until all Coomassie has been eluted from pieces. After cooling samples to 4°C, scintillation fluid is added, and radioactivity measured in a liquid scintillation counter. Since low levels of radioactivity (below 1000 dpm) are incorporated into specific proteins, it is not reliable enough to just count protein after organic extraction (varying levels of labeled lipid remain despite these extractions). The specific proteins, therefore, have to be identified and counted separately.

COMMENTS

The procedure given works only satisfactory if special care is taken about the quality of both acceptor preparations and microsomes. As will be seen from the results obtained with incubations No. 6 and 7, heat treatment prevents any incorporation of [^{14}C]palmitic acid from the palmitoyl-CoA precursor. Expecially, the microsomes are most sensitive to temperature, and lose acylation activity even after short periods at room temperature. Also, the detergents used for virus solubilization seem crucial. Whilst NP40 and Triton X-100 yield positive results, viral acceptor proteins prepared in octyl-β-D-glucoside fail to incorporate fatty acid. It has also been noted that the acyl donor [^{14}C]plamitoyl CoA is subject to degradation when microsomes are present. If this happens too rapidly with a given enzyme source, it may be useful to use alternative substrates. [^{3}H]fatty acids have been applied with some success, but they only work if incubation mixtures also contain at least 0.05 mM ATP (Berger and Schmidt 1984). Some recent results indicate that supplementing this latter mixture with CoA and MgCl (both at similar concentration as ATP) can yield higher levels of [^{3}H]acyl CoA and also of [^{3}H]acyl protein (Schmidt and Burns, unpublished). Instead of using exogenous acceptor, we also studied palmitoylation *in vitro* of endogenous acceptor proteins (Mack et al. 1987). In this case, microsomes from virus infcted cells (VSV in mouse myeloma cells) were incubated with [^{14}C]palmitoyl CoA, and [^{14}C]fatty acids were transferred into both viral G-protein and unidentified host cell proteins. Further development and optimization of the cell free systems for palmitoylation of proteins may facilitate the identification and isolation of a protein palmitoyltransferase.

Acknowledgments. We thank Rudi Rott for continuous support and encouragement in our studies on acylation. All efforts by Eva Kroell and Margot Seitz with editorial work and typing of the manuscript are gratefully acknowledged. Work herein has been supported by Sonderforschungsbereich 47 of the Deutsche Forschungsgemeinschaft, and research initation grant MB 173 of Kuwait University. This work is presented in partial fulfilment of the requirements for Dr. rer. nat. of Marion Schmidt at the FB 15 of Justus-Liebig-Universität, Giessen, Federal Republic of Germany.

REFERENCES

Berger M, Schmidt MFG (1984) J Biol Chem 259:7245-7252
Berger M, Schmidt MFG (1986) J Biol Chem 261:14912-14918
Mack D, Berger M, Schmidt MFG, Kruppa J (1987) J Biol Chem 262:4297-4302
Magee AI, Courtneidge SA (1985) EMBO J 4:1137-1144
McIlhinney RAJ, Pelley SJ, Chadwick JK, Cowley GP (1985) EMBO J 4:1145-1152
Schmidt MFG (1982) Virology 116:327-338
Schmidt MFG (1983) Curr Top Microbiol Immunol 102:101-130
Sefton BM, Buss JE (1987) J Cell Biol 104:1449-1453

Lipid-Mediated Protein Glycosylation: Assembly of Lipid-Linked Oligosaccharides and Post-Translational Oligosaccharide Trimming

W. McDowell* and R.T. Schwarz[+]

*MRC Collaborative Centre, 1-3 Burtonhole Lane, Mill Hill,
London NW7 1AD, U.K.
[+]Laboratoire de Chimie Biologique, Université des Sciences et Techniques
de Lille-Flandres-Artois, F-59655 Villeneuve d'Ascq Cédex, France

INTRODUCTION

The complex and high mannose oligosaccharide side-chains of asparagine-linked glycoproteins are derived from a common precursor molecule, $Glc_3Man_9(GlcNAc)_2$ (Robbins et al. 1977) which is assembled on a dolichol-pyrophosphate (Dol-PP) carrier in the rough endoplasmic reticulum (RER) (Kornfeld and Kornfeld 1985). Assembly is achieved through the sequential addition of GlcNAc, Man and Glc residues, either by direct transfer from sugar nucleotide donors, or via dolichol monophosphate (Dol-P) sugar intermediates (Kornfeld and Kornfeld 1985). The first reaction in the pathway is the addition of GlcNAc-1-P to Dol-P forming Dol-PP-GlcNAc. Transfer of GlcNAc completes the assembly of the di-N-acetylchitobiosyl core (Hubbard and Ivatt 1981). The first five mannose residues are transferred directly from GDP-Man forming the $Man_5(GlcNAc)_2$-PP-Dol intermediate (Hubbard and Ivatt 1981) which is the substrate for the transfer of mannose residues from Dol-P-Man (Chapman et al. 1980) forming the $Man_9(GlcNAc)_2$-PP-Dol intermediate. The three glucose residues are all transferred from Dol-P-Glc (Staneloni et al. 1980; Murphy and Spiro 1981) forming the major lipid-linked oligosaccharide, $Glc_3Man_9(GlcNAc)_2$-PP-Dol.

En bloc transfer of the fully assembled oligosaccharide $Glc_3Man_9(GlcNAc)_2$ to protein occurs co-translationally in the lumen of the RER (Kornfeld and Kornfeld 1985), and once protein-bound, the oligosaccharide is subjected to a series of trimming reactions in which the three glucose residues and four mannose residues are removed (Kornfeld and Kornfeld 1985). Incomplete removal of the mannose residues yields high mannose structures. Once the

---

Abbreviations. Glc, glucose; Man, mannose; GlcNAc, N-acetylglucosamine; GDP, guanosine diphosphate; UDP, uridine diphosphate; Dol, dolichol.

molecule has been trimmed to the $Man_5(GlcNAc)_2$ stage, a GlcNAc residue is added and two further mannose residues are removed. Sequential addition of GlcNAc, galactose, fucose, and neuraminic acid residues completes the synthesis of the complex type side-chains (Kornfeld and Kornfeld 1985).

Glucose removal is catalyzed by glucosidases I and II (Grinna and Robbins 1979; Burns and Touster 1982) which are resident in the RER (Kornfeld and Kornfeld 1985). A RER alpha-mannosidase may remove one mannose residue (Bischoff and Kornfeld 1983) before the glycoprotein is translocated to the cis compartment of the Golgi apparatus where alpha-mannosidase I completes mannose trimming to the $Man_5(GlcNAc)_2$ trimming intermediate (Tabas and Kornfeld 1979). This oligosaccharide is the substrate for N-acetylglucosaminyl-transferase I (Harpaz and Schachter 1980) which resides in the medial compartment of the Golgi apparatus (Kornfeld and Kornfeld 1985) as does Golgi alpha-mannosidase II which catalyzes the final step in mannose trimming (Tulsiani et al. 1982a). The glycosyltransferases which catalyze complex oligosaccharide formation, are located in the trans Golgi compartment (Kornfeld and Kornfeld 1985).

Specific inhibitors of the assembly of the lipid-linked oligosaccharide precursor (Schwarz and Datema 1982a) and the trimming glycosidases (Schwarz and Datema 1984) have proved to be most useful in investigating the pathways of lipid-mediated protein glycosylation. Furthermore, these inhibitors have been used as tools for studying the biological significance of protein-bound carbohydrate and cell biological phenomena such as intracellular transport, secretion, and the multiplication of enveloped viruses (Hubbard and Ivatt 1981; Schwarz and Datem 1982a, 1984; Kornfeld and Kornfeld 1985).

Several inhibitors of the dolichol pathway have been described, and their mechanism of inhibition elucidated. Tunicamycin which blocks the first step in the dolichol pathway (Lehle and Tanner 1976; Schwarz and Datema 1982a) is certainly the best well-known and most frequently used inhibitor of this group. Analogues of glucose and mannose containing deoxy- or fluoro-substituents on carbons 2, 3, 4, or 6 are also useful inhibitors of lipid-linked oligosaccharide assembly (Schwarz and Datema 1982a; McDowell et al. 1985; McDowell et al. 1987a). They are metabolized *in vivo* to their respective GDP and/or UDP derivatives (Schwarz and Datema 1982a) which are the actual inhibitory agents, and interfere with steps further along the pathway involving mannosyl- and glucosyltransferases (Schwarz and Datema 1982a; McDowell et al. 1985; McDowell et al. 1987a). Glucosamine differs from the other sugar analogues in that the underivatized sugar is the inhibitory agent (Koch et al. 1979; McDowell et al. 1986), and the inhibition can be rapidly reversed (Schwarz and Datema 1982a). UDP-glucosamine is also an inhibitor of lipid-linked saccharide assembly (McDowell et al. 1986), but it is only formed in liver cells metabolizing galactosamine (McDowell et al. 1986).

Inhibitors of oligosaccharide processing are currently in vogue for studying the relative importance of high mannose and complex oligosaccharidic side-chains of asparagine-linked glycoproteins as well as intracellular transport and the assembly of enveloped viruses (Schwarz and Datema 1984). Several inhibitors of the trimming glycosidases are now commercially available. These include the glucosidase inhibitors bromoconduritol (glucosidase II), 1-deoxynojirimycin (glucosidases I and II), and castanospermine (glucosidase I), and the mannosidase inhibitors 1-deoxymannojirimycin (mannosidase I), and swainsonine (mannosidase II). The modes of action of many of the trimming inhibitors were elucidated by observing their effect on the glycosylation of the membrane glycoproteins of enveloped viruses. Some interesting biological phenomena have been found as a consequence of these studies. Thus, the formation of some influenza A viruses (Romero et al. 1983; Datema et al. 1984b; Elbein et al. 1984) and Rous sarcoma virus (Bosch and Schwarz 1984; Bosch et al. 1985) is not affected by either glucosidase or mannosidase inhibitors with the exception of the influenza A virus, fowl plague virus, the haemagglutinin of which is metabolically unstable when equipped with oligosaccharides of the composition $Glc_1Man_9(GlcNAc)_2$ in the presence of the glucosidase inhibitor bromoconduritol (Datema et al. 1984b). Observations made with sindbis virus (Schlesinger et al. 1985; McDowell et al. 1987b), vesicular stomatitis virus (Schlesinger et al. 1984), and mouse hepatitis virus (Repp et al. 1985) have indicated an essential role for glucose trimming in the establishment of a functional conformation for some viral glycoproteins. The initial stages of mannosidase trimming also appear to be important in the sindbis virus-baby hamster kidney cell system for determining the final destination of the viral glycoproteins (McDowell et al. 1987b) since treatment with the mannosidase I inhibitor 1-deoxymannojirimycin resulted in budding of the virus from intracellular membranes (McDowell et al. 1987b).

The aim of this chapter is to present some of the methods in use in our laboratory for the radioactive labeling, isolation, and characterization of oligosaccharides linked to both lipid and protein. In addition, the use of specific inhibitors of protein glycosylation will be presented.

MATERIALS AND EQUIPMENT

In vivo **labeling**

D-[2-^{3}H]mannose 18.3 Ci/mmol; D-[6-^{3}H]glucosamine hydrochloride 40 Ci/mmol;
D-[6-^{3}H]galactose 28 Ci/mmol (Amersham International)
Glass petri dishes 9 cm diameter

Chloroform - methanol (2:1 by volume)

$37^{\circ}C$ incubator with a 5% CO_2 atmosphere

Refrigerated laboratory centrifuge (Christ)

3 ml glass homogenizer (Braun)

Lipid extraction

Chloroform - methanol (2:1 by volume, CM)

Chloroform - methanol - water (10:10:3 by volume, CMW)

4 mM $MgCl_2$

Theoretical upper phase: methanol - water - 1 M $MgCl_2$ - chloroform

(96:94:0.8:6 by volume)

Methanol

Refrigerated labortory centrifuge (Christ)

Preparation of chicken embryo cell microsomes

Plastic petri dishes 15 cm diameter

Buffered EDTA (10 mM EDTA, 140 mM NaCl, 7 mM KCl, 1 mM potassium phosphate, pH 7.0)

Rubber policeman

Refrigerated laboratory centrifuge

Buffers: 20 mM Tris-HCl, 150 mM NaCl, pH 7.5; 20 mM Tris-HCl, pH 7.5;

20 mM Tris-HCl, 100 mM NaCl, 0.4 mM $MgCl_2$, 0.4 mM $MnCl_2$, pH 7.5

30 mM $MnCl_2$ and 100 mM NaCl

Dounce homogenizer

Beckman refrigerated ultracentrifuge (L2-55 or larger) with Ti50 rotor

In vitro **biosynthesis of lipid-linked saccharides**

Chicken embryo cell microsomal membrane preparation

Thermostated water bath

GDP-[U-^{14}C]mannose 300 Ci/mol; UDP-N-acetyl-[U-^{14}C]glucosamine 230 Ci/mol;

UDP-[U-^{14}C]glucose 300 Ci/mol (Amersham International)

GDP-Man, UDP-Glc, UDP-GlcNAc (Boehringer Mannheim)

Dolichol phosphate (Sigma)

Triton X-100

Analysis of lipid-linked saccharides

Rotary evaporator (Büchi Instruments)

DEAE cellulose (Whatman DE 52)

Glass columns 1 x 5 cm

Glacial acetic acid

99% methanol

n-butanol

0.2 M and 0.4 M ammonium acetate in 99% methanol

Chloroform - methanol - water (10:10:3 by volume, CMW)

0.12 M ammonium acetate in CMW

Silica G-60 thin layer plates (Merck)

Thin layer chromatogram scanner (Berthold LB 2842)

1 M HCl and n-propanol

Thermostated water bath

Endoglucosaminidase H (Boehringer Mannheim or Genzyme)

Endo H buffer double strength: 43 mM citric acid, 114 mM sodium phosphate, pH 5.5

2% sodium azide

Pronase (Serva)

Pronase buffer 10 x strength: 1.5 M Tris-HCl, 15 mM $CaCl_2$, pH 7.8

Gel filtration

BioGel P4 for separation of oligosaccharides (BioRad)

BioGel P6 for separation of glycopeptides (BioRad)

Glass columns 1 x 150 cm and 1 x 100 cm

0.02% sodium azide

0.1 M pyridinium acetate

Blue dextran (Pharmacia) and phenol red (Merck)

Bovine serum albumin fraction V (Serva)

Fraction collector which can hold about 160-200 small tubes of 2.5 ml volume

Spectrophotometer with UV lamp

Glycosidase digestion

Alpha-mannosidase (Sigma)

Mannosidase buffer: 200 mM acetate, 40 μM $ZnCl_2$, pH 5.0

Two glass columns 1 x 5 cm

Dowex 50W-X8 and Dowex 1-X2 ion exchange resins (BioRad)

Cellulose thin layer plate (Merck)

Rat liver microsome preparation with alpha-glucosidase activity

0.2 M sucrose; 100 mM sodium phosphate buffer, pH 6,9; Sorvåll refrigerated centrifuge with SS42 rotor; Beckman ultracentrifuge with Ti50 rotor; glass-teflon homogenizer

Glycosylation inhibitors

2-deoxy-D-glucose and 2-deoxy-2-fluoro-D-glucose (Calbiochem or Sigma)

D-Glucosamine (Serva)

2-deoxy-2-fluoro-D-mannose (Calbiochem, not listed anymore in the 1987 catalogue)

4-deoxy-D-mannose and 4-deoxy-4-fluoro-D-mannose (commercially not available, can be synthesized chemically, see Rasmussen et al. 1983)

GDP and UDP derivatives of the above mannose and glucose analogues are not available commercially and must be synthesized chemically by coupling the sugar-1-phosphate to GDP- or UDP-morpholidate (Elbein 1966; MacDonald 1966; Moffatt 1966)

Tunicamycin (Boehringer Mannheim, Calbiochem, Sigma)

Inhibitors of oligosaccharide trimming

1-deoxynojirimycin, castanospermine, bromoconduritol, swainsonine, and 1-deoxymannojirimycin (Boehringer Mannheim, Calbiochem, Genzyme)

EXPERIMENTAL PROCEDURES

Labeling and extraction of lipid-linked oligosaccharides and glycoproteins from cell cultures

Tissue cultured cells are grown on glass petri dishes with a diameter of 9 cm. Remove medium, and wash three times with warm (37°C) phosphate buffered saline, and add 2.5 ml of warm Earles medium containing 10 mM sodium pyruvate, or 10 mM glucose, or 10 mM fructose.

If the cells have been virus infected, the radiolabeled sugar is added 4 h after infection. Otherwise, the label can be added after changing the medium. Lipid-linked oligosaccharides can be labeled by adding 100 μCi of either D-[2-^{3}H]mannose, D-[6-^{3}H]glucosamine, or D-[6-^{3}H]galactose to the cell incubation medium. The oligosaccharides are thus labeled in their Man, GlcNAc, or Glc residues, respectively. Incubation of cells with the labeled sugar for 2 h is normally long enough to sufficiently label both the lipid-linked oligosaccharides and the protein-bound oligosaccharides. The major lipid-linked oligosaccharide observed under these conditions, is $Glc_3Man_9(GlcNAc)_2$-PP-Dol. If lipid-linked oligosaccharides smaller than this are to be observed, then shorter incubation times (3 min or less) must be employed, and it is advisable to use 200 μCi of labeled sugar.

After labeling, the radioactive medium is removed by aspiration, and the plate immediately flooded with 4 ml ice-cold chloroform - methanol (2:1 by volume; CM). The cells are scraped together and the plate rinsed with 1 ml CM. They are then homogenized in CM in a ground glass homogenizer, and centrifuged at 4000 g for 5 min. The cell pellet is resuspended in 2 ml CM with sonication, and centrifuged again.

The two supernatant fluids (CM extracts) are combined and Folch-washed to remove water-soluble radioactive components as follows:

a) Add 1/5 volume of 4 mM $MgCl_2$ and mix well.

b) Centrifuge to give two phases. Remove upper phase and discard.

c) Rinse the tube twice with 1 ml of an artificial upper phase without mixing. Each time, remove the upper phase and discard.

d) Add 1 ml of theoretical upper phase, and thoroughly mix the contents of the tube.

e) Centrifuge and remove the upper phase.

f) The washed lower phase constitutes the CM extract, and contains Dol-P-monosaccharides and Dol-PP-oligosaccharides with up to five glycosyl residues (Dol-PP-$(GlcNAc)_2$ Man_3). An aliquot can be taken for liquid scintillation counting. Allow CM to evaporate before adding scintillation fluid.

After extraction with CM, the cell residue is washed with 200 μl of methanol, and centrifuged at 4000 g for 5 min. The methanol is removed by aspiration, the pellet is washed four times with 4 ml of water by sonication, and centrifuged at 4000 g for 10 min. The pellet is washed once more with 200 μl of methanol and centrifuged.

The pellet is now extracted twice with 2 ml of chloroform - methanol - water (10:10:3 by volume, CMW), and once with 1 ml of CMW.

The resulting extract is the CMW extract and contains lipid-linked oligosaccharides. An aliquot can be taken for liquid scintillation counting. Allow CMW to evaporate before adding scintillation fluid.

The lipid-free residue containing glycoproteins, can be dissolved in 1 ml of 1% sodium dodecyl sulphate at $100^{\circ}C$ for 15 min for liquid scintillation counting, or polyacrylamide electrophoresis sample buffer for gel electrophoresis, or in pronase buffer for pronase digestion to generate glycopeptides.

Preparation of a membraneous enzyme fraction suitable for the *in vitro* assembly of lipid-linked saccharides

The medium is removed from 15 confluent monolayers of chicken embryo cells grown on plastic petri dishes (diameter 15 cm), and the cells washed once with 5 ml buffered EDTA solution (10 mM EDTA, 140 mM NaCl, 7 mM KCl, 1 mM potassium phosphate, pH 7.0).

The cells are incubated with 2.5 ml per plate of buffered EDTA at $37^{\circ}C$ for 20 min. They are then scraped together with a rubber policeman, and collected by centrifuging at 1000 g at room temperature for 5 min. All subsequent steps are performed at $4^{\circ}C$.

The pelleted cells are washed once by gently resuspending in 5 ml of 20 mM Tris-HCl, 150 mM NaCl, pH 7.5, and centrifuged at 1000 g at $4^{\circ}C$ for 5 min. The pellet is resuspended in approximately 2 ml of the same buffer, and is then added in drops to 10 volumes of ice-cold 20 mM Tris-HCl, pH 7.5. The cells are allowed to swell in the buffer for 20 min, and are then homogenized with 12 strokes in a Dounce glass homogenizer. 0.1 volume of 30 mM $MgCl_2$ and 100 mM NaCl are added, and three more strokes are given.

The homogenate is centrifuged at 1000 g at $4^{\circ}C$ for 3 min to pellet cell debris and nuclei. The supernatant fluid is taken and centrifuged at 100,000 g at $4^{\circ}C$ for 60 min (Beckman Ti50 rotor).

The resulting membraneous pellet is suspended in 1250 μl of 20 mM Tris-HCl, 100 mM NaCl, 0.4 mM $MgCl_2$, 0.4 mM $MnCl_2$, pH 7.5, to give a protein concentration of 15-20 mg/ ml. This membraneous preparation can be used immediately for assaying the biosynthesis of dolichol-linked saccharides.

In vitro biosynthesis of lipid-linked saccharides

The microsomal membrane fraction prepared as described above is used. The final concentrations of the buffer components are 13 mM Tris-HCl, pH 7.5, 100 mM NaCl, 0.3 mM $MgCl_2$, 0.3 mM $MnCl_2$ in a total volume of 120 μl.

Dol-P may be added as a suspension in 0.2% Triton X-100. The final concentrations of Dol-P and Triton X-100 in the assay should be 0.3 mg/ml and 0.017%, respectively.

All incubations are carried out at $37^\circ C$, and are stopped by adding 2 ml ice-cold CM followed by rapidly mixing with a long Pasteur pipette to disperse the denatured reaction mixture. The mixture is centrifuged at 4000 g for 10 min to pellet the protein which is extracted once more with 2 ml CM. The CM extracts are combined and Folch-washed as described for the *in vivo* studies above.

The pellet is washed with methanol and water, and extracted with CMW essentially as described above with the exception that 1 ml of water is used three times for the wash.

Lipid-linked saccharides can be synthesized *in vitro* at $37^\circ C$ as follows:

a) $([^{14}C]GlcNAc)_2$-PP-Dol: 0.05 μCi UDP-$[^{14}C]GlcNAc$ are incubated for 20 min.

b) $[^{14}C]Man$-P-Dol: 0.05 μCi GDP-$[^{14}C]Man$ are incubated for 5 min.

c) $[^{14}C]Glc$-P-Dol: 0.05 μCi UDP-$[^{14}C]Glc$ are incubated for 15 min.

d) $([^{14}C]Man)_9(GlcNAc)_2$-PP-Dol: Incubate with 22 μM UDP-GlcNAc for 20 min to form a pool of Dol-PP-$(GlcNAc)_2$. Then incubate with 0.05 μCi GDP-$[^{14}C]Man$ for 10 min.

e) $Man_9([^{14}C]GlcNAc)_2$-PP-Dol: Incubate with 0.05 μCi UDP-$[^{14}C]GlcNAc$ for 20 min. Then incubate with 8.3 μM GDP-Man for 20 min.

f) $Glc_3Man_9([^{14}C]GlcNAc)_2$-PP-Dol: Incubate as described in e), but 5 μM UDP-Glc is included with the labeled GDP-Man.

g) $([^{14}C]Glc)_3Man_9(GlcNAc)_2$-PP-Dol: Incubate with 22 μM UDP-GlcNAc for 20 min. Then incubate with 8.3 μM GDP-Man and 0.05 μCi UDP-$[^{14}C]Glc$ for 15 min.

h) $([^{14}C]Man)_5(GlcNAc)_2$-PP-Dol: Incubate with 22 μM UDP-GlcNAc for 20 min. Then add EDTA (inhibits the subsequent formation of Man-P-Dol) to give a final concentration of 10 mM, and after 5 min, add 0.05 μCi GDP-$[^{14}C]Man$. Incubate for 60 min.

The products of reactions a) to c) are soluble in CM, and those of reactions d) to h) in CMW. In these latter reactions, it is preferable to examine also the CM extract if intermediates are being looked for.

Analysis of lipid-linked saccharides

The methods described are equally applicable to lipid-linked oligosaccharides prepared *in vivo* and *in vitro*.

1. Ion exchange chromatography on DEAE cellulose (acetate form)

This is used to purify the negative charged lipid-linked saccharides from non-polar lipids, and to separate Dol-PP-oligosaccharides from Dol-P-monosaccharides. It must be used to isolate the Dol-P and Dol-PP derivatives from the CM extracts prepared from *in vivo* incubations, before further analysis on these derivatives can be carried out.

a) Preparation of DEAE cellulose in the acetate form:

Gently stir up the DEAE cellulose (Whatman DE-52) in 3 volumes of 1 M NaOH for 30 min. Collect on a sintered glass filter.Wash on the filter with water to neutrality. Wash with 4 volumes of ethanol and 4 volumes of methanol. Allow to dry and break up clumps. Stir overnight in glacial acetic acid. Collect by filtration and wash with 99% methanol. Store in 99% methanol at 4°C.

b) Isolation of Dol-P derivatives:

Prepare a column (1 x 2 cm) of DEAE cellulose in the acetate form equilibrated with 16 ml of 99% methanol. Sample to be analyzed is dried using rotary evaporation (use nitrogen gas to break the vacuum). Dissolve in 1 ml n-butanol and apply to the DEAE cellulose column. Neutral compounds are eluted with 16 ml of 99% methanol. Dol-P-sugars are eluted with 8 ml of 0.2 M ammonium acetate in 99% methanol. Dol-PP- oligosaccharides may be eluted with 8 ml of 0.4 M ammonium acetate in 99% methanol.

c) Isolation of Dol-PP-oligosaccharides:

A column (1 x 2 cm) of DEAE cellulose in the acetate form equilibrated with 50 ml of CMW is used preferably for the separation of Dol-PP-oligosaccharides from other glycolipids. Samples in 3 ml of CMW are applied to the column. Impurities and Dol-P-sugars are eluted with 100 ml of equilibrating solvent. Dol-PP-oligosaccharides are eluted with 50 ml of 0.12 M ammonium acetate in CMW.

Ammonium acetate present in the eluates must be removed by phase partitioning prior to further analysis: To eluates containing 99% methanol, add 2 volumes of chloroform, mix and centrifuge to give two phases, and remove upper phase. Wash with theoretical upper phase as described above. Lower phase contains the desalted lipids. To eluates containing CMW, add 20 ml of chloroform and 6 ml of water, mix and centrifuge to give two phases, and remove upper phase. Wash with theoretical upper phase as described above. Lower phase contains the desalted lipids.

2. Thin layer chromatography

This is performed on plates of silica G-60 (Merck) which have been washed with methanol, dried at room temperature, activated at 100°C, and stored under vacuum over phosphorous pentoxide.

To identify Dol-P-sugars and to separate Dol-PP-oligosaccharides containing up to 5 glycosyl residues, chloroform - methanol - ammonia - water (65:35:4:4 by volume) is used.

To identify and separate Dol-PP-GlcNAc and Dol-PP-(GlcNAc)$_2$, chloroform - methanol - water (60:39:6 by volume) is used.

Radioactivity on the plates can be measured either by scraping portions of the silica into scintillation vials, adding scintillation fluid and counting, or by scanning with a commercial thin-layer chromatogram scanner (e.g., Berthold LB 2842 linear analyzer with computerized peak integration).

3. Mild acid hydrolysis

Saccharides linked by mono- or pyrophosphate bonds to dolichol are labile to acid. The oligosaccharides present in lipid-linked oligosaccharides can be conveniantly analyzed by gel filtration (see below) after their release by mild acid hydrolysis.

The sample to be hydrolyzed is dried by rotary evaporation. 0.5 ml of 2 M HCl, and 0.5 ml of n-propanol are added and then vortexed. Hydrolysis is accomplished by heating in a water bath at 50°C for 15 min.

After cooling, the sample is rotary evaporated to remove the HCl and n-propanol. 1 ml of chloroform and 1 ml of water are added, mixed by vortexing, and centrifuged to give two phases. The upper phase containing the released oligosaccharides is rotary evaporated and taken up in 50 μl of water.

4. Digestion with Endoglucosaminidase H

Oligosaccharides released from their lipid carriers by mild acid hydrolysis (see above) can be treated with Endoglucosaminidase H (Endo H) to facilitate their subsequent analysis by gel filtration (see below). Endo H treatment is also used to release high mannose oligosaccharides from glycopeptides prepared by pronase digestion (see below) of the lipid-free residue from *in vivo* cell labeling experiments.

Add 50 μl of double strength buffer comprising 43 mM citric acid, 114 mM sodium phosphate, pH 5.5, to the mild acid hydrolyzed sample (in 50 μl of water). 2 μl of 2% sodium azide and 2.5 mU Endo H are added and incubated at 37°C for 18 h. The reaction can be stopped by heating in a boiling waterbath for 5 min.

5. Pronase digestion of glycoproteins

The lipid-free pellet obtained after the extraction of cells with CM and CMW is suspended with sonication in 170 μl of water.

20 μl of buffer comprising 1.5 M Tris-HCl, pH 7.8, 15 mM $CaCl_2$ and 10 μl of pronase (10mg/ml predigested at 37°C for 1 h) are added. 1 drop of toluene is added, and the sample is incubated at 37°C. After 24 and 48 h, respectively, another 10 μl of pronase are added.

After 72 h, the reaction is stopped by heating the samples in a boiling water bath for 10 min. Precipitated protein is removed by centrifugation at 4000 g for 5 min, and the supernatant fluid containing the glycopeptides is kept for further analysis.

6. Gel filtration of glycopeptides on BioGel P6

Glycopeptides prepared by pronase digestion can be separated by gel filtration on a column (1 x 100 cm) of BioGel P6 (200-400 mesh) equilibrated with 0.1 M pyridinium acetate, pH 5.0. This procedure also desalts the pronase digest enabling the release of high mannose oligosaccharides by Endo H treatment (see above).

The pronase digest containing blue dextran and phenol red as markers for the void and inclusion volumes, respectively, in a volume not exceeding 1 ml, is applied to the column. Elute with 0.1 M pyridinium acetate, pH 5.0, and fractions of 0.5 ml are collected. As this is usually a preparative procedure, aliquots of 20 μl of each fraction are taken for liquid scintillation counting.

The fractions containing the glycopeptides are pooled, rotary evaporated, digested with Endo H and subjected to gel filtration on BioGel P4.

7. Gel filtration of oligosaccharides on BioGel P4

Oligosaccharides released from lipid-linked oligosaccharides, or from glycopeptides by Endo H treatment, can be separated on columns (1 x 150 cm) of BioGel P4 (-400 mesh).

The sample containing 4 mg of bovine serum albumin (BSA) as a void volume marker, is loaded onto the column in a volume not exceeding 200 μl. Elute with water containing 0.02% sodium azide, and fractions of 0.4 ml are collected. The void volume of the column is determined by measuring the absorbance at 280 nm caused by the presence of BSA in those fractions. Radioactivity in each fraction is determined by liquid scintillation counting.

The column can be calibrated with oligosaccharides of known composition prepared by the *in vitro* incubations described above.

8. Glycosidase digestion of oligosaccharides

a) Digestion with *alpha-mannosidase* is useful in the characterization of lipid-derived oligosaccharides and the high mannose oligosaccharides derived from glycopeptides.

The sample is taken up in 120 μl of 200 mM acetate buffer, pH 5.0, containing 40 μM of $ZnCl_2$ and 0.02% sodium azide. 1 Unit of Jack Bean alpha-mannosidase is added. The enzyme must be dialyzed against the above buffer before use, and is stable for 4 weeks if stored at 4^oC. After 14 and 20 h, respectively, of incubation at 37^oC, a further 0.5 Units are added.

After 24 h, the reaction is stopped by heating in a boiling waterbath for 10 min. The boiled sample is diluted with 1 ml of water, and deionized by passage through coupled columns (1 x 1 cm) of Dowex 50W-X8 (H^+) and Dowex 1-X2 ($HCOO^-$). Deionized material is eluted with 20 ml of water and rotary evaporated. Dried material is taken up in 50 μl of water.

The liberated mannose can be quantitated after gel filtration on BioGel P4 or after cellulose thin-layer chromatography using pyridine - ethyl acetate - acetic acid - water (36:36:8:20 by volume) as a solvent.

b) Digestion with *alpha-glucosidase* is useful to determine whether or not glucose is present in oligosaccharides which have, for example, been synthesized *in vivo* in the presence of glucosidase trimming inhibitors.

The sample is split into aliquots containing 2000 cpm radioactive label. Digestion is carried out in 100 mM phosphate buffer, pH 6.9, containing 7.5 mM EDTA and 0.1% Triton X-100. 10 μl of a rat liver microsomal alpha-glucosidase preparation (see below) is added to give a final volume of 30 μl. Incubate at 37^{0}C for 2-3 h. The reaction is stopped by heating in a boiling waterbath for 5 min.

Denatured protein is removed by centrifugation, and the supernatant fluid is kept for analysis by gel filtration on BioGel P4. [^{14}C] labeled Glc$_3$Man$_9$GlcNAc and Man$_9$GlcNAc are added to the samples as reference compounds prior to gel filtration.

Preparation of rat liver microsomes with alpha-glucosidase activity

All centrifugation steps are at 4^{0}C, and the sucrose and buffer solutions should be ice-cold.

Two rat livers are quickly excised, weighed, and placed in 50 ml of ice-cold 0.25 M sucrose. They are cut into small pieces with scissors and a scalpel, and the volume of sucrose adjusted to 5 ml/g liver. Homogenize the liver in a glass-teflon homogenizer by hand until a good suspension is obtained. Centrifuge in a refrigerated laboratory centrifuge at 2700 g for 20 min.

Take the supernatant fluid and centrifuge in a Sorvall SS34 rotor at 12,000 g for 20 min. Take the supernatant fluid and centrifuge in a Beckman Ti50 rotor at 100,00 g for 60 min. The pellet is taken up in 100 mM sodium phosphate buffer, pH 6.9 (approximately 3 ml per liver).

The microsomal preparation can be stored at -70^{0}C in small portions (50-100 μl), and can be frozen and thawed twice without affecting the alpha-glucosidase activity.

Inhibition of lipid-linked oligosaccharide formation *in vivo*

In vivo inhibitors of protein glycosylation are listed in Table 1 together with an indication of the effective concentration.

Stock solutions (100 mM) of the fluoro and deoxy sugars in water, and of tunicamycin (1 mg/ml in 0.1 M NaOH) are stable when stored at -20^{0}C.

Glucosamine (0.5 M) adjusted to pH 7.5 with 1 M NaOH, is prepared fresh each time and used immediately.

Table 1

Glycosylation inhibitor	Reaction(s) inhibited / Comments	Inhibitory concentration	References
2-Deoxy-D-glucose (2dGlc)	Formation of Man-P-Dol, Glc-P-Dol, and GlcNAc-PP-Dol due to formation of 2dGlc-P-Dol. Mannosylation of $(GlcNAc)_2$-PP-Dol and Man$(GlcNAc)_2$-PP-Dol due to the formation of dGlc$(GlcNAc)_2$-PP-Dol and dGlcMan$(GlcNAc)_2$-PP-Dol. The glucosylation of Man$_9(GlcNAc)_2$-PP-Dol.	2 mM (pyruvate-containing medium) or 5 mM (5 mM Glc-containing medium)	Schwarz and Datema (1982a)
2-Deoxy-2-fluoro-D-glucose (2FGlc)	Formation of Man-P-Dol and Glc-P-Dol. Pool size of UDP-Glc reduced. Glc$_3$Man$_5(GlcNAc)_2$-PP-Dol is the major lipid-linked oligosaccharide formed.	0.5 mM	Schwarz and Datema (1982a)
2-Deoxy-2-fluoro-D-mannose (2FMan)	Formation of Man-P-Dol. Mannosylation of $(GlcNAc)_2$-PP-Dol. Guanosine (1 mM) must be present in the medium to prevent depletion of intracellular guanosine pools which could lead to inhibition of protein synthesis.	0.2 mM	McDowell et al. (1985)
4-Deoxy-4-fluoro-D-mannose (4FMan)	Mannosylation of Man$(GlcNAc)_2$-PP-Dol. 4FMan-P-Dol synthesis detected.	1-10 mM (0.55 mM Glc-containing medium)	McDowell et al. (1987a)

Table 1 (continued)

Glycosylation inhibitor	Reaction(s) inhibited / Comments	Inhibitory concentration	References
4-Deoxy-D-mannose (4dMan)	Formation of Man-P-Dol and Glc-P-Dol. Major lipid-linked oligosaccharide is $Man_9(GlcNAc)_2$-PP-Dol. 4dMan may be incorporated into lipid-linked oligosaccharide.	10 mM (0.55 mM Glc-containing medium)	McDowell et al. (1987a)
D-Glucosamine (GlcN)	Lipid-linked oligosaccharide assembly. $Man_3(GlcNAc)_2$-PP-Dol accumulates. Glucose-containing media must be used. GlcN present at the same concentration as glucose in the medium. Inhibition reversible (within 20 min) after removal of the GlcN-containing medium.	5-10 mM (5-10 mM Glc-medium)	Schwarz and Datema (1982a) Elbein (1983)
Tunicamycin	Formation of GlcNAc-PP-Dol. 1 mg/ml stock solution in 0.1 M NaOH	1-4 μg/ml	Schwarz and Datema (1982a) Lehle and Tanner (1976)

Table 2

Inhibitor	Reaction(s) inhibited / Comments	Inhibitory concentration	References
GDP-2dGlc	Man-P-Dol, Glc-P-Dol, GlcNAc-PP-Dol synthesis due to 2dGlc-P-Dol formation. Formation of 2dGlc(GlcNAc)$_2$-PP-Dol.	100 μM	Schwarz and Datema (1982a) Schwarz and Datema (1982b)
UDP-2-dGlc	Glc-P-Dol synthesis.	120 μM	Schwarz and Datema (1982a) Schwarz and Datema (1982b)
GDP-2FMan	Man-P-Dol, Glc-P-Dol, GlcNAc-PP-Dol synthesis due to 2FMan-P-Dol formation. Mannosylation of (GlcNAc)$_2$-PP-Dol.	150 μM	McDowell et al. (1985)
UDP-2FMan	Glc-P-Dol synthesis.	100 μM	McDowell et al. (1985)
GDP-2FGlc	No effect on lipid-linked saccharide formation. Retinol phosphate mannose synthesis in rat liver microsomes inhibited.	400 μM	Datema and Schwarz (1984a)
UDP-2FGlc	Glc-P-Dol synthesis.	200 μM	Schwarz and Datema (1982a) Schwarz and Datema (1982b)

Table 2 (continued)

Inhibitor	Reaction(s) inhibited / Comments	Inhibitory concentration	References
GDP-4FMan	Mannosylation of Man(GlcNAc)$_2$-PP-Dol.	150 μM	McDowell et al. (1987a)
GDP-4dMan	Man-P-Dol, Glc-P-Dol, GlcNAc-PP-Dol synthesis due to 4dMan-P-Dol formation.	150 μM	McDowell et al. (1987a)
Tunicamycin	Formation of GlcNAc-PP-Dol. Mannosyl transfer not affected. Glc-P-Dol synthesis at high concentrations (5 μg/ml).	0.5 μg/ml	Schwarz and Datema (1982a) Lehle and Tanner (1976) Schwarz and Datema (1982b)

Table 3

Trimming inhibitor	Concentration	Enzyme inhibited	Oligosaccharide composition	References
N-Methyl-1-deoxynojirimycin	1 mM	Glucosidase I + II	Glc$_3$Man$_{7,8,9}$(GlcNAc)$_2$	Romero et al. (1983)
1-Deoxynojirimycin	1-5 mM	Glucosidase I + II	Glc$_{1-3}$Man$_{7-9}$(GlcNAc)$_2$	Saunier et al. (1982)
Castanospermin	52 μM	Glucosidase I	Glc$_3$Man$_9$(GlcNAc)$_2$	Pan et al. (1983)
Bromoconduritol	2.4 mM	Glucosidase II	Glc$_1$Man$_{7,8,9}$(GlcNAc)$_2$	Datema et al. (1984b)
1-Deoxymanno-jirimycin	2 mM	Mannosidase I	Man$_9$(GlcNAc)$_2$	Elbein et al. (1984) McDowell et al. (1987b)
Swainsonine	3 μM	Mannosidase II	Hybrid-type	Tulsiani et al. (1982b)

Inhibition of lipid-linked saccharide formation *in vitro*

The nucleoside diphosphate esters of the sugar analogues described above and tunicamycin can be used to inhibit lipid-linked saccharide formation in microsomal membrane preparations as indicated in Table 2.

Inhibitors of oligosaccharide trimming *in vivo*

The conditions for each inhibitor should be worked out for each cell type used. Conditions used in this laboratory are indicated in Table 3.

In our laboratory, we use glucosidase inhibitors with fructose-containing media to prevent interference with the uptake of the inhibitor by the greater amount of carbon source present in the medium. Thus, if glucose is present in the medium, it may reduce the uptake of the glucosidase inhibitors because they will probably use the same transporter to enter the cell. Alternatively, low glucose-containing media may be used.

Some of these trimming inhibitors may also inhibit the assembly of lipid-linked oligosaccharide. Therefore, it is advisable to check for this possibility in the cell type being used.

Acknowledgments. The authors' work herein has been supported by Sonderforschungsbereich 47 of the Deutsche Forschungsgemeinschaft, and the Fonds der Chemischen Industrie. For part of the work, R.T. Schwarz had received a Heisenberg Stipendium of the Deutsche Forschungsgemeinschaft. The constant support of Drs R. Rott, C. Scholtissek, and J. Montreuil is gratefully acknowledged.

REFERENCES

Bischoff J, Kornfeld R (1983) J Biol Chem 258:7907-7910
Bosch JV, Schwarz RT (1984) Virology 132:95-109
Bosch JV, Tlusty A, McDowell W, Legler G, Schwarz RT (1985) Virology 143:342-346
Burns DM, Touster O (1982) J Biol Chem 257:9991-10000
Chapman A, Fujimoto K, Kornfeld S (1980) J Biol Chem 255:4441-4446
Datema R, Schwarz RT (1984a) Biosci Rep 4:213-221
Datema R, Romero PA, Rott R, Schwarz RT (1984b) Arch Virol 81:25-39
Elbein AD (1966) In: Neufeld EF, Ginsburg V (eds) Methods in Enzymology, Vol 8, Academic Press New York London, pp 142-145
Elbein AD (1983) In: Fleischer S, Fleischer B (eds) Methods in Enzymology, Vol. 98, Academic Press New York London, pp 135-154
Elbein AD, Legler G, Tlusty A, McDowell W, Schwarz RT (1984) Arch Biochem Biophys 235:579-588
Grinna LS, Robbins PW (1979) J Biol Chem 254:8814-8818
Harpaz N, Schachter H (1980) J Biol Chem 255:4894-4902

Hubbard SC, Ivatt RJ (1981) Ann Rev Biochem 50:555-583

Koch HU, Scholtissek C, Schwarz RT (1979) Eur J Biochem 94:515-522

Kornfeld R, Kornfeld S (1985) Ann Rev Biochem 54:631-664

Lehle L, Tanner W (1976) FEBS Lett 71:167-170

MacDonald DL (1966) In: Neufeld EF, Ginsburg V (eds) Methods in Enzymology, Vol 8, Academic Press New York London, pp 121-125

McDowell W, Datema R, Romero PA, Schwarz RT (1985) Biochemistry 24:8145-8152

McDowell W, Weckbecker G, Keppler DOR, Schwarz RT (1986) Biochem J 233:749-754

McDowell W, Grier TJ, Rasmussen JR, Schwarz RT (1987a) Biochem J 248:523-531

McDowell W, Romero PA, Datema R, Schwarz RT (1987b) Virology 161:37-44

Moffat JG (1966) In: Neufeld EF, Ginsburg V (eds) Methods in Enzymology, Vol 8, Academic Press New York London, pp 136-142

Murphy LA, Spiro RG (1981) J Biol Chem 256:7487-7494

Pan YT, Hori H, Saul R, Sanford BA, Molyneux RJ, Elbein AD (1983) Biochemistry 22:3975-3984

Rasmussen JR, Tafuri SR, Smale ST (1983) Carbohydr Res 116:21-29

Repp R, Tamura T, Boschek CB, Wege H, Schwarz RT, Niemann H (1985) J Biol Chem 260:15873-15879

Robbins PW, Hubbard SC, Turco SJ, Wirth DF (1977) Cell 12:893-900

Romero PA, Datema R, Schwarz RT (1983) Virology 130:238-242

Saunier B, Kilker RD, Tkacz JS, Quaroni A, Herscovics A (1982) J Biol Chem 257:14155-14161

Schlesinger S, Malfer C, Schlesinger MJ (1984) J Biol Chem 259:7597-7601

Schlesinger S, Koyama AH, Malfer C, Gee SL, Schlesinger MJ (1985) Virus Res 2:139-149

Schwarz RT, Datema R (1982a) Adv Carbohydr Chem Biochem 40:287-379

Schwarz RT, Datema R (1982b) In: Ginsburg V (ed) Methods in Enzymology, Vol 83, Academic Press New York London, pp 432-443

Schwarz RT, Datema R (1984) TIBS 9:32-34

Staneloni RJ, Ugalde R, Leloir LF (1980) Eur J Biochem 105:275-278

Tabas I, Kornfeld S (1979) J Biol Chem 254:11655-11663

Tulsiani DRP, Hubbard SC, Robbins PW, Touster O (1982a) J Biol Chem 257:3660-3668

Tulsiani DRP, Harris TM, Touster O (1982b) J Biol Chem 257:7936-7939

Immunochemical Characterization of the Ia Associated Invariant Chain

N. Koch

Institut für Immunologie und Genetik, Deutsches Krebsforschungszentrum,
Im Neuenheimer Feld 280, Postfach 10194, D-6900 Heidelberg, Federal Republic of Germany

INTRODUCTION

Ia antigens are intracellularly associated with a non-polymorphic invariant polypeptide (invariant chain or Ii). Both, Ia and Ii are coexpressed in macrophages, B cells, and in some activated T cells. Treatment of cells with gamma interferon leads to coregulation of Ia and Ii by induction/enhancement of transcription.

During intracellular transport, the invariant chain is post-translationally modified. Very early after membrane insertion, two high mannose carbohydrate side chains are linked to asparagin residues of the invariant chain. Transport to the Golgi leads to addition of palmitic acid to a sole cysteine adjacent to the cytoplasmic part of the membrane. In the Golgi, the invariant chain is O-glycosylated, and the high mannose carbohydrates are trimmed. In trans-Golgi compartments, sialic acids are linked to N- and O-linked carbohydrates. A small portion of the invariant chain is found to be modified to a proteoglycan and possibly appears at the cell surface.

EXPERIMENTAL PROCEDURES

Production of an antiserum against the invariant chain

The low amount of invariant chain produced by eukaryotic cells, makes it difficult to isolate sufficient amounts of antigen for immunisation of rabbits. Therefore, we have chosen recombinant DNA technology to produce large amounts of antigen (Koch et al. 1987). A 0.79 kb fragment of the invariant chain cDNA was cloned into a prokaryotic expression vector (PEX in Figure 1).

The cloning site which is available in all three reading frames, is located at the 3' end of the MS2 polymerase gene. This construct permits expression of a fusion protein containing a molecular weight of about 10,000 of MS2 polymerase, and a molecular weight of about 20,000 of the invariant chain. Large quantities of the fusion protein were obtained from overnight cultures of bacteria containing the construct (yield: about 1 mg of fusion protein/ml of bacterial culture). The fusion protein was enriched to high purity by several extraction steps with urea. The purified fusion protein was used for immunizing a rabbit (500 μg of fusion protein mixed with complete Freund's adjuvant, and four weeks later restimulation with the same dose of fusion protein in incomplete Freund's adjuvant). Two weeks later, the rabbit was bled, and the serum was used for immunochemistry.

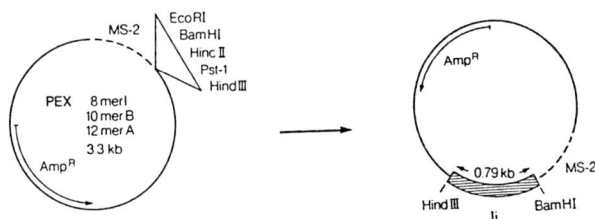

Fig. 1

Metabolic labeling of cells (Kvist et al. 1982)

B lymphocytes and macrophages which express the invariant chain, were used for immunochemical studies. Cells were cultured in RPMI-1640 complemented with 10% FCS. Cells were taken from logarithmic growth to achieve a high rate of protein synthesis, and washed twice with medium depleted from the amino acid which was used for labeling. 10^7 cells were suspended in medium containing 200 μCi of the labeled amino acid and 10% of dialyzed FCS. In order to obtain a good uptake of the radioactive amino acid into the cells, for the first 5 min of labeling, the volume was kept small (200 μl/10^7 cells). Then, 10 ml of prewarmed medium were added. The labeling procedure was stopped by addition of cold medium (4°C) and centrifugation. The cells now either were lysed with NP40 (see below), or the cell pellet was frozen at -20°C (never freeze the lysate because the proteins become insoluble). Labeling with [³H]palmitic acid follows the same procedure, except that 500 μCi/10^7 cells were used. The [³H]palmitic acid is purchased in ethanol. Since ethanol is toxic for cells, the dilution of ethanol - medium must not be higher than 3%. In general, a good labeling will be obtained within 1 h. Longer incubation times may lead to interconversion of the isotopes into other compounds.

Immunoprecipitation (Koch and Hammerling 1986)

The cell pellet was resuspended in 500 μl PBS with 10 mM Tris, pH 7.4, containing protease inhibitors - Trasylol (1:1000), and 1 mM phenylmethylsulfonyl fluoride. Then, 50 μl of
10% NP40 was added for lysis of the cells. DNA and debris was removed by high speed centrifugation at 10,000 x g for 2 min. The supernatant was preabsorbed with a suspension of 50
μl of protein A sepharose for 2 h. Immunoprecipitation was performed by adding 5 μl of protein A sepharose suspension, 5 μl of serum, or 10 μg of monoclonal antibody to the pre-
cleared lysate. After rotating over night at 4°C, the immunoprecipitates were washed three
times with 0.25% NP40 containing protease inhibitors. A syringe with a small needle is useful to separate the supernatant from the sepharose beads. Then, the immunoprecipitates were
treated with sample buffer (see below) and subjected to electrophoresis.

Two-dimensional gel electrophoresis (O'Farrel et al. 1977)

1. <u>First dimension</u>

   Non-equilibrium pH gradient electrophoresis (NEPHGE):

   The immunoprecipitates were resuspended in NEPHGE sample buffer and incubated at
   room temperature for 1 h. Do not heat the samples to avoid carbamylation which leads to ar-
   tificially charged polypeptides. Cylindrical NEPHGE gels which have been prepared one day
   in advance (see below), were loaded with the samples and overlaid with 0.01 M phosphoric
   acid (anode positive) (0.02 M sodium hydroxide at the bottom). Run at 550 V for 5 h (constant voltage). Then, the gels were carefully removed from the glass tubes, and incubated in
   SDS sample buffer at room temperature for 1 h.

2. <u>Second dimension</u>

   SDS-polyacrylamide gel electrophoresis (SDS-PAGE):

   The rod gels were now fixed on the top of the second dimensional gel with hot agarose
   (1% agarose in SDS sample buffer).

3. <u>Preparation of the NEPHGE gels</u>

   The glass tubes must be very clean. Treat them with chromic acid and ethanol/NaOH,
   and wash carefully with distilled water. The tubes were capped at the bottom with parafilm,
   and filled with the reaction mixture using a syringe with a long needle.

Reaction mixture:

5.5 g urea - 1.3 ml 30% acrylamide/bis (17:1)

2 ml Triton X-100

2 ml double distilled water

0.5 ml Ampholine, pH 3.5-10.0

Dissolve the urea before starting the reaction with 14 μl 10% AP and 9 μl TEMED. After filling the tubes, the reaction mixture is overlaid with 20 μl of distilled water.

REFERENCES

Koch N, Hammerling GJ (1986) J Biol Chem 261:3434-3440
Koch S, Schulz A, Koch N (1987) J Immunol Meth 103:211-220
Kvist S, Wiman K, Claesson L, Peterson PA, Dobberstein B (1982) Cell 29:61-69
O'Farrel PZ, Goodman HM, O'Farrel PH (1977) Cell 12:1133-1142

Acylation and Activation of Venom Phospholipase A$_2$ Enzymes by Acyl Imidazolides

A.J. Lawrence and S. Chettibi

Department of Cell Biology, University of Glasgow, Glasgow G12 8QQ, Scotland, U.K.

INTRODUCTION

Phospholipase A$_2$ enzymes hydrolyze phosphatidyl phospholipids to lysophosphatides and free fatty acid with very little specificity, either for head group type, or acyl chain substituents. They are present in all eukaryotic cells, and catalyze the rate-limiting step in lipid remodeling (deacylation and reacylation) which may regulate membrane fluidity, or remove partially oxidized acyl groups, and it also provides a major route for the release of arachidonate in the inflammatory response. Phospholipase A$_2$ activity can be enhanced by calcium, or by the phosphorylation of an associated inhibitory protein (lipocortin) (Flower and Blackwell 1979). The enzyme A$_2$ is present in high concentrations in most animal venoms and in pancreatic secretion. Many primary sequences are known, and some crystal data is available (Brunie et al. 1985). There is a high degree of homology among the enzymes from diverse sources (Dufton and Hider 1983), and they share common regulatory features, and appear to have similar active sites. Venom enzymes are not normally directly lytic, and in most cases, they are secreted with direct lytic factors which also synergize the catalytic action. Both, lysophosphatide and fatty acid reaction products can be powerful activators in *in vitro* systems, and the generally accepted mechanism invokes changes in the phospholipid substrate surface produced by "detergent" action. Little evidence exists to support specific activation of the enzyme by either species except in the case of bee venom phospholipase A$_2$. This enzyme can be irreversibly activated by weakly activated derivatives of long-chain fatty acids which rapidly add a single acyl group to the protein, and abolishes the reversible activation by fatty acids (Drainas and Lawrence 1978). This is evidence for an allosteric site operated either by fatty acids, or by acylation; occupation of the site greatly increases the stability of the protein to proteases and thiol agents suggesting that it causes a major change conformation (Camero Diaz et al. 1985). The putative activation site has high reactivity towards long-chain weak acylating agents (particularly acyl imidazolides) determined by the acyl group. The biological role of activation is unclear, but it is not restricted to the honey bee enzyme. As yet, there is no evidence for *in vivo* acylation of these enzymes. This paper describes a general method for acylation of susceptible enzymes together with assays for activation and stabilisation.

EXPERIMENTAL PROCEDURES

Enzymes and reagents

a) Phospholipase A_2 can be activated in whole venoms or the purified form. These include purified enzymes from honey bee, *Naja mocambique*, and *Vipera russelli*. The enzyme has also been activated in whole venoms from honey bee and a variety of bumble bees. A single venom sac provides adequate material for activation.

b) Acyl imidazolides are prepared by adding a slight excess of carbonyldiimidazole (Sigma) to fatty acid in dry acetone. Purification of the derivative is not necessary, but is readily achieved by evaporating to dryness extracting in petroleum ether, evaporating, and redissolving in dry acetone. Stock solutions are made at 2% w/v, and diluted in dry acetone before use. Long-chain acyl imidazoides are hydrolyzed very slowly at neutral pH.

c) Activation requires a pH above 6.5, and buffer compounds with free primary amino groups should be avoided. Triethanolamin, MOPS, or bicine buffers are all satisfactory.

Assays

We use conductimetric methods for all assays (Lawrence et al. 1974; Drainas et al. 1981), but activation can also be determined using an erythrocyte lysis assay (Drainas and Lawrence 1980) requiring, at minimum, simple visual observation. The conductivity cell is of the type shown, and is connected to a sensitive, linear conductivity meter. Conventional methods for phospholipase A_2 assays, but with the same substrates and reaction conditions, should give similar results.

Solution for activity assays:
    10 mM triethanolamine/Cl, pH 8.0, with 0.1 mM dinonanoyl or dioctanoyl phosphatidylcholine

Solutions for activation assays:
a) Erythrocyte leakage assay:
    120 mM sucrose

10 mM MOPS/Na, pH 7.4

0.01 mM bovine serum albumin

Add 20 μl of a 30% v/v suspension of rabbit erythrocytes in isotonic saline medium just prior to use.

b) Egg lecithin assay:

10 mM triethanolamine/Cl, pH 8.0

20% v/v propan-1-ol

0.1 mM calcium chloride

1 mg/ml purified egg phosphatidylcholine

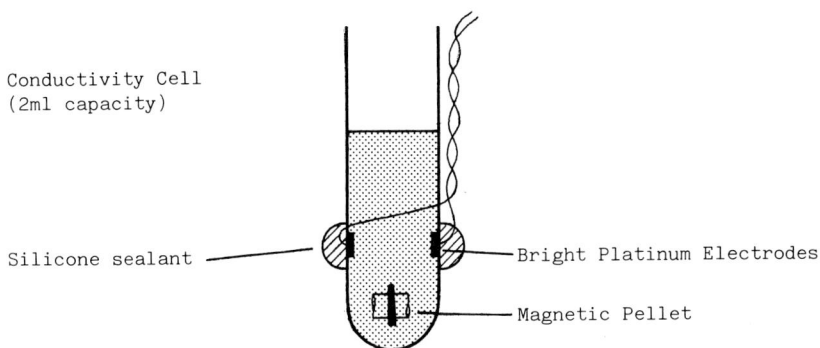

Conductivity Cell
(2ml capacity)

Silicone sealant

Bright Platinum Electrodes

Magnetic Pellet

Fig. 1

Enzyme activation

Bee venom phospholipase A_2 is incubated at approximately 1 mg/ml in 100 mM buffer, and activation started by adding a molar equivalent of acyl imidazolide in acetone. For the pure enzyme, 0.1 ml of 1 mg/ml protein (66 nmol) requires 1.8 μg of myristoyl imidazolide (MW 278), and it is convenient to add 1.8 μl of 1 mg/ml sub-stock solution in acetone.

All experiments described use 2 ml of the assay medium at $37^{\circ}C$ with 2 μl enzyme samples. In the erythrocyte-leakage test, the untreated enzyme shows a weak response at this concentration, but activation produces a family of sigmoidal curves with increasingly steep slopes. The maximum slope is the most convenient parameter for estimation of activation. Activation factors vary with the calcium concentration of the medium; with a background level of approximately 10 μM (as here), factors of 50-fold are typical, but this could fall to about 10-fold at 1 mM calcium concentration. Using a semi-quantitative lysis assay, the non-lytic enzyme becomes highly lytic as activation proceeds. These results can be substantiated using the egg lecithin hydrolysis assay. The treatment has minimal effect on the rate at which the enzyme attacks dinonanoyl phosphatidylcholine (approx. 20% activation). In consequence, this assay is highly suitable for stability studies, and allows the enhanced stability to a variety of inactivating conditions to be measured directly.

COMMENTS

Albumin mediated leakage of erythrocytes is very sensitive to acylation of phospholipase A_2, and insensitive to fatty acid stimulation of the enzyme because free fatty acids bind to albumin more strongly than they do to phospholipase A_2. The assay is, therefore, specific for activation by acylation. The biological implication is that where activation appears to have greatest potential for increasing the toxic activity of the enzyme, free fatty acids cannot act. Fatty acids shorter than myristic acid, are very weak activators (measured in the absence of albumin), but detectable activation is observed with the bound heptanoyl residue. Myristoyl imidazolide gives both the fastest rate of activation and the highest activation factor in the erythrocyte assay, although the oleoyl residue was more effective in the egg phosphatidylcholine assay (Camero Diaz et al. 1985). The technique described, allows a variety of substituted acyl groups to be added to the protein including branched and ring compounds. A proximal linear sequence of at least 4 carbon (or equivalent) atoms seems to be necessary for rapid activation.

REFERENCES

Brunie S, Bolin J, Gewirth D, Sigler PB (1985) J Biol Chem 260:9742-9749
Camero Diaz RE, Elansari O, Lawrence AJ, Lyall F, McLeod WA (1985) Biochim Biophys Acta 830:52-58
Drainas D, Lawrence AJ (1978) Eur J Biochem 91:131-138
Drainas D, Lawrence AJ (1980) FEBS Lett 114:93-97
Drainas D, Harvey E, Lawrence AJ, Thomas A (1981) Eur J Biochem 114:239-245
Dufton MJ, Hider RC (1983) Eur J Biochem 137:545-551
Flower RJ, Blackwell GJ (1979) Nature 278:456-459
Lawrence AJ, Moores GR, Steele J (1974) Eur J Biochem 48:277-286

Isolation and Characterization of Bovine Erythrocyte Acetylcholinesterase Solubilized by PI-PLC

H. Ikezawa and R. Taguchi

Faculty of Pharmaceutical Sciences, Nagoya City University,
3-1 Tanabe-dori, Mizuho-ku, Nagoya 467, Japan

INTRODUCTION

By use of bacterial phosphatidylinositol-specific phospholipase C (PI-PLC), several enzymes such as alkaline phosphatase (Ikezawa et al. 1976), 5'-nucleotidase (Low and Finean 1978; Taguchi and Ikezawa 1978), alkaline phosphodiesterase I (Nakabayashi and Ikezawa 1984; Nakabayashi and Ikezawa 1986), acetylcholinesterase (Low and Finean 1977; Futerman et al. 1983; Taguchi et al. 1984), trehalase (Takesue et al. 1986), nucleoside diphosphatase (Sawaki et al. 1983), and lysosomal acid ATPase (Maeda et al. 1986) have been reported to be released or converted to hydrophilic, smaller forms from the plasma membranes and ER membranes of mammalian cells, synaptosomal membrane of *Torpedo* electric organ, and lysosomal membrane of chicken liver. Especially, acetylcholinesterase was effectively liberated from *Torpedo* electric organ (Futerman et al. 1985; Low et al. 1987), and bovine erythrocytes (Taguchi et al. 1984; Taguchi and Ikezawa 1987), and established as PI-anchoring enzyme. The aims of this experiment are to isolate bovine erythrocyte acetylcholinesterase, and to confirm the presence of inositol-containing links in the enzyme molecule.

EQUIPMENT AND SOLUTIONS

1. Purification of bovine erythrocyte acetylcholinesterase

Purified PI-PLC of *Bacillus thuringiensis*: 1 unit/ml 150 mM NaCl, 5 mM Na-phosphate, pH 7.4

Bovine erythrocyte suspension: 10% (v/v) in 150 mM NaCl, 5 mM Na-phosphate, pH 7.4

10 mM Tris-HCl, pH 7.5

NaCl solution: 0.1 M in 10 mM Tris-HCl, pH 7.5; 0.2 M in 10 mM Tris-HCl, pH 7.5; 0.2 M in 10 mM Na-phosphate, pH 7.5

10 mM edrophonium chloride ("Tensilon", Kyorin Pharmaceutical Co. Ltd.) in 0.2 M NaCl; 10 mM Na-phosphate, pH 7.5

Means for concentration and ultrafiltration: PM-10 membrane (Amicon); Centriflo CF-25 (Amicon)

DEAE-cellulose (James River Corporation, Berlin), column size 3.2 x 50 cm

Affinity gel prepared from CNBr-activated Sepharose 4B (Pharmacia) by the method of Berman (1973)

Spacer atom and ligand part: $-[NH-(CH_2)_3-NH-(CH_2)_3-NH-CO-(CH_2)_2-CO]_2-NH-\text{⊙}-N^+(CH_3)_2$, column size 2.6 x 10 cm

Sephadex G-75 (Pharmacia), column size 2.6 x 100 cm

Sepharose 6 B (Pharmacia), column size 2.6 x 100 cm

2. GC-MS analysis of myo-inositol

Acetylcholinesterase in 6 N HCl (200 μg/ml), and

6 N HCl containing bovine serum albumin and myo-inositol (250 μg and 250 ng/ml).

In these HCl solutions, aqueous solutions of samples are diluted with conc. HCl.

Pyridine

Trimethylchlorosilane

Hexamethyldisilazane

Glass ampules

Rotary pump

Glass column 2 m x 2 mm packed with 3% Silicon SE 30 on Uniport HP 60/80 mesh

GC-MS equipment: JEOL DX 300 Mass Spectrometry fitted with JMA-3500 Mass Data Analysis System

3. Spectrophotometric measurement of acetylcholinesterase activity

Enzyme solution

10 mM Na-phosphate, pH 7.5

10 mM DTNB: 10 mM 5,5'-dithiobis-2-nitrobenzoic acid in 10 mM Na-phosphate, pH 7.5

12.5 mM acetylthiocholine chloride

Reading Spectrophotometer (Shimadzu double-beam UV-180)

EXPERIMENTAL PROCEDURES

1. Flow sheet of the purification

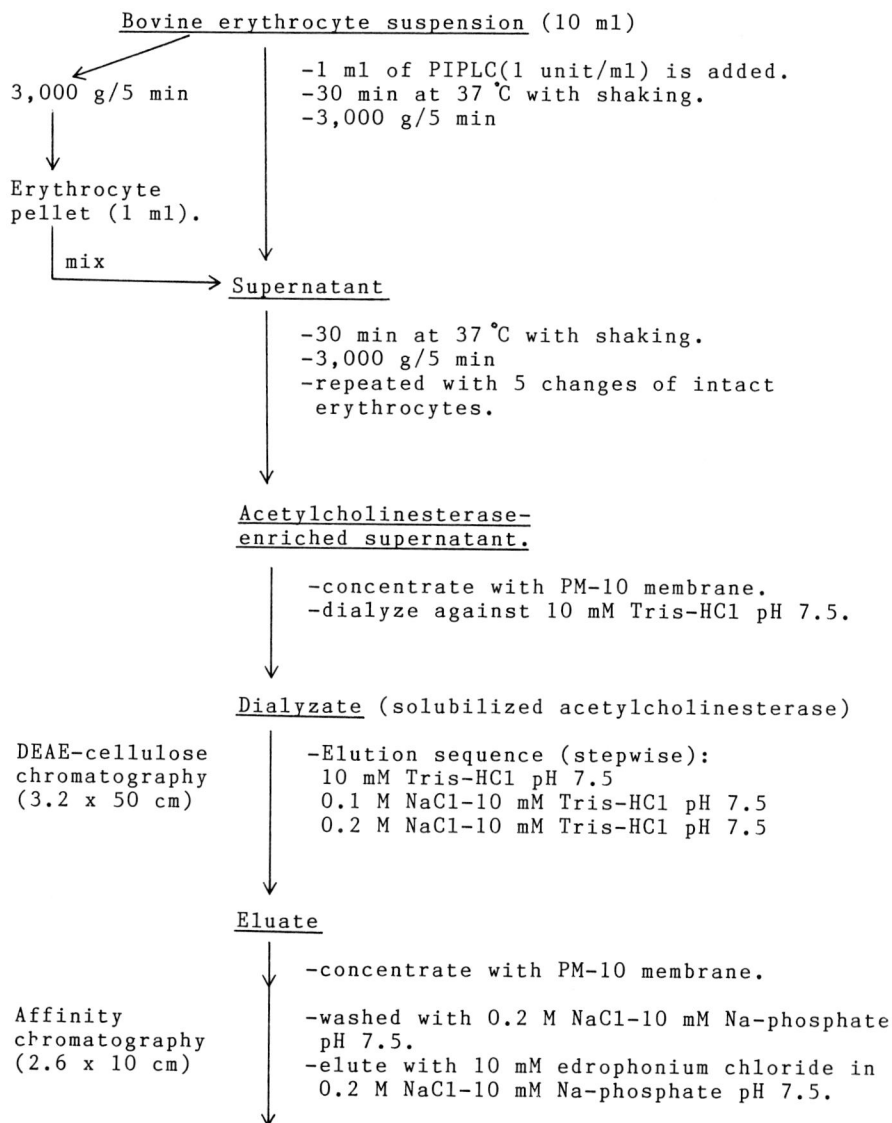

Bovine erythrocyte suspension (10 ml)

3,000 g/5 min

-1 ml of PIPLC(1 unit/ml) is added.
-30 min at 37 °C with shaking.
-3,000 g/5 min

Erythrocyte
pellet (1 ml).

mix

Supernatant

-30 min at 37 °C with shaking.
-3,000 g/5 min
-repeated with 5 changes of intact
 erythrocytes.

Acetylcholinesterase-
enriched supernatant.

-concentrate with PM-10 membrane.
-dialyze against 10 mM Tris-HCl pH 7.5.

Dialyzate (solubilized acetylcholinesterase)

DEAE-cellulose
chromatography
(3.2 x 50 cm)

-Elution sequence (stepwise):
 10 mM Tris-HCl pH 7.5
 0.1 M NaCl-10 mM Tris-HCl pH 7.5
 0.2 M NaCl-10 mM Tris-HCl pH 7.5

Eluate

-concentrate with PM-10 membrane.

Affinity
chromatography
(2.6 x 10 cm)

-washed with 0.2 M NaCl-10 mM Na-phosphate
 pH 7.5.
-elute with 10 mM edrophonium chloride in
 0.2 M NaCl-10 mM Na-phosphate pH 7.5.

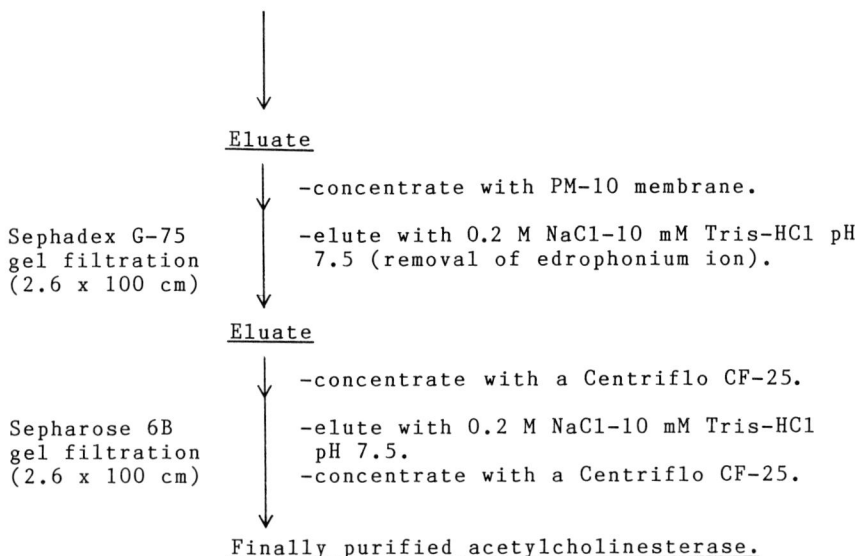

Eluate

    —concentrate with PM-10 membrane.

Sephadex G-75   —elute with 0.2 M NaCl-10 mM Tris-HCl pH
gel filtration   7.5 (removal of edrophonium ion).
(2.6 x 100 cm)

Eluate

    —concentrate with a Centriflo CF-25.

Sepharose 6B   —elute with 0.2 M NaCl-10 mM Tris-HCl
gel filtration   pH 7.5.
(2.6 x 100 cm)   —concentrate with a Centriflo CF-25.

Finally purified acetylcholinesterase.

2. GC-MS analysis of myo-inositol in the molecule of acetylcholinesterase

a) In a glass ampule, add 1 ml acetylcholinesterase in 6 N HCl (200 μg/ml). Seal, and heat
   at 110°C for 40 h. (In the control run, 1 ml of 6 N HCl containing 250 μg bovine serum
   albumin and 250 ng myo-inositol is treated in the same way.)

b) Evaporate at 37°C under N_2 stream, and dry completely by vacuum evaporation with a
   rotary pump.

c) Add 50 μl pyridine, 5 μl trimethylchlorosilane, and 10 μl hexamethyldisilazane. Mix and
   stand for 10 min.

d) Withdraw 10 μl aliquots, and inject into a glass column packed with 3% Silicon SE 30 on
   Uniport HP 60/80 mesh, and set in GC-MS equipment. Temperature program at 170°-
   250°C at 4°C/min.

e) Measure mass-spectric fragment of myo-inositol hexatrimethylsilyl ether at m/z 612, 507,
   432, and 305.

3. Measurement of acetylcholinesterase activity (Ellman et al. 1961)

Mix 0.1 ml enzyme with 0.8 ml 10 mM Na-phosphate, pH 7.5, 50 μl 10 mM DTNB, and 50
μl 12.5 mM acetylthiocholine chloride in 1 ml cuvette. Follow the reaction spectrophotometrically at 412 nm at 25°C by Reading Spectrophotometer (Shimadzu, double-beam UV-180).

COMMENTS

Bovine erythrocyte acetylcholinesterase purified by this procedure has a specific activity of approximately 5,200 μmol acetylthiocholine hydrolyzed/min/mg protein. Polyacrylamide gel electrophoresis (7.5% gel at pH 9.4) shows the single band (Taguchi et al. 1984).

REFERENCES

Berman JD (1973) Biochemistry 12:1710-1715
Ellman GL, Courtney KD, Andres V Jr, Featherstone RM (1961) Biochem Pharmacol 7:88-95
Futerman AH, Low MG, Silman I (1983) Neurosci Lett 40:85-89
Futerman AH, Low MG, Ackermann KE, Sherman WR, Silman I (1985) Biochem Biophys Res Commun 129:312-317
Ikezawa H, Yamanegi M, Taguchi R, Miyashita T, Ohyabu T (1976) Biochim Biophys Acta 450:154-164
Low MG, Finean JB (1977) FEBS Lett 82:143-146
Low MG, Finean JB (1978) Biochim Biophys Acta 508:565-570
Low MG, Futerman AH, Ackermann KE, Sherman WR, Silman I (1987) Biochem J 241:615-619
Maeda K, Ohta Y, Nakabayashi T, Taguchi R, Ikezawa H (1986) Biochem Int 12:855-863
Nakabayashi T, Ikezawa H (1984) Cell Struct Funct 9:247-263
Nakabayashi T, Ikezawa H (1986) J Biochem 99:703-712
Sawaki K, Taguchi R, Ikezawa H (1983) Chem Pharm Bull 31:2769-2778
Taguchi R, Ikezawa H (1978) Arch Biochem Biophys 186:196-201
Taguchi R, Suzuki K, Nakabayashi T, Ikezawa H (1984) J Biochem 96:437-446
Taguchi R, Ikezawa H (1987) J Biochem 102:803-811
Takesue Y, Yokota K, Nishi Y, Taguchi R, Ikezawa H (1986) FEBS Lett 201:5-8

Characterization of Amphiphilic Forms of Cholinesterases by their Interactions with Non-Denaturing Detergents in Centrifugation and Charge-Shift Electrophoresis

J. Massoulié[*,1], J.P. Toutant[+], and S. Bon[*]

[*]*Laboratoire de Neurobiologie, Ecole Normale Supérieure, 46 rue d'Ulm,
F-75230 Paris Cédex 05, France*
[+]*Station de Physiologie Animale, Institut National de la Recherche Agronomique,
9 place Pierre Viala, F-34060 Montpellier Cédex, France*

INTRODUCTION

Vertebrates possess two distinct cholinesterases: acetylcholinesterase (AChE, EC 3117), and butyrylcholinesterase (BuChE, EC 3118) which differ in their substrate and inhibitor specificities (Augustinsson 1963; review in Silver 1974). Both enzymes may exist in several molecular forms which correspond to different quaternary associations of catalytic subunits: three globular forms corresponding to the monomer, dimer, and tetramer (G_1, G_2, and G_4), and three asymmetric forms in which one, two, or three tetramers are associated with a collagen-like element (A_4, A_8, and A_{12} (for review see Massoulié and Bon 1982)).

The central nervous system of insects constitutes a very rich source of AChE with no BuChE (Arpagaus and Toutant 1985; Melanson et al. 1985; Gnagey et al. 1987). This enzyme exists only in the forms of globular monomers (G_1) and dimers (G_2) with no asymmetric form (Forunier et al. 1987; Gnagey et al. 1987; Toutant et al. 1988, in press).

In vertebrates and invertebrates, the globular forms of cholinesterases exist as amphiphilic molecules which bind non-denaturing detergents such as Triton X-100 and hydrophilic (or non-amphiphilic) molecules which do not interact with non-denaturing detergents. This distinction coincides only partially with that based upon the requirement of detergent for solubilization ("membrane-bound" and "soluble" fractions).

---

[1] To whom correspondence should be addressed.

Abbreviations. AChE, acetylcholinesterase; BuChE, butyrylcholinesterase; Brij 96, polyoxyethylene(10)oleylether; ChE, cholinesterase; CTAB, cetyltrimethylammonium; DOC, sodium deoxycholate; DS, detergent-soluble; LSS, low salt-soluble; PI-PLC, phosphatidylinositol phospholipase C; TX100, Triton X-100, polyoxyethylene(9-10)octylphenol.

Key words. Acetylcholinesterase; butyrylcholinesterase; amphiphilic; charge-shift electrophoresis; detergents.

The amphiphilic character is operationally defined by changes in the hydrodynamic properties of the molecules which accompany the binding of the detergent: the Stokes radius determined by molecular sieve chromatography is increased (Bon and Massoulié 1980; Grassi et al. 1982; Rosenberry and Scoggin 1984); the sedimentation coefficient in sucrose gradient centrifugation is decreased (Bon and Massoulie 1980; Grassi et al. 1982); the rate of electrophoretic migration is decreased with increasing concentrations of TX100 (Borsum and Bjerrum 1986). Positive and negative charged detergents can also modify the electrophoretic migrations of amphiphilic molecules (Helenius and Simons 1977; Arpagaus and Toutant 1985; Stieger and Brodbeck 1985). Finally, amphiphilic molecules of AChE may be incorporated into phospholipid vesicles (Römer-Lüthi et al. 1980; Ott and Brodbeck 1984). Non-amphiphilic molecules are defined in a negative manner by the absence of the above properties.

Readers interested in detailed reviews of the interactions between amphiphilic AChE and lipids, may refer to Ott (1985), Brodbeck (1986), and Massoulié and Toutant (in press). In the present paper, we describe two methodological approaches for the characterization of amphiphilic cholinesterases using sucrose gradient centrifugation and charge-shift electrophoresis.

EXPERIMENTAL PROCEDURES

Ultracentrifugation in sucrose gradients

Futerman et al. (1983, 1984) reported that dimers of *Torpedo* AChE sedimented at 4.8-5.0 S when solubilized and analyzed in the non-denaturing detergent Brij 96. This enzyme sedimented at 6.0-6.3 S when it was extracted and analyzed in the presence of TX100 (see also Bon and Massoulié 1980), and at 7.4 S in the presence of cholate. Considering these differences, we systematically examined the influence of the two detergents TX100 and Brij 96 on the sedimentation of amphiphilic forms of AChE.

1. Low salt extract of rabbit muscle in the absence of detergent

Three identical samples of a low salt extract of denervated *semimembranosus* muscle of adult rabbit were centrifuged in three types of gradients: without detergent, with 1% TX100, or with 0.5% Brij 96. Figure 1 shows that two major forms sedimented at 9.4 S (G_4) and 4.0 S (G_1) in the absence of detergent. In the presence of 1% TX100, the G_4 and G_1 forms sedimented at 9.4 S and 3.0 S, respectively, and a shoulder of the 3.0 S peak was observed around 4.5 S. In the presence of 0.5% Brij 96, the sedimentation of the G_4 form was unchanged (9.4 S), but two separate peaks were observed at 3.0 S and 1.0 S corresponding to G_2 and G_1 forms. We conclude that the G_4 form is a non-amphiphilic, hydrophilic form, whereas both G_1 and

134

G_2 are amphiphilic forms, although they are solubilized in the absence of detergent. The binding of Brij 96 decreased the sedimentation of the G_1 and G_2 forms more markedly than that of TX100, and allowed their separation in the gradient indicating a difference in the size and/or the density of the detergent micelles which associated with their hydrophobic domains.

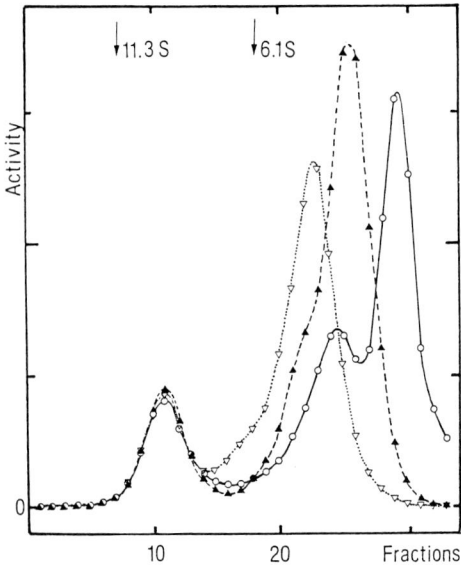

Fig. 1. Analysis of AChE globular forms in a low salt soluble traction of rabbit *semimembranosus* muscle. 5-20% sucrose gradient containing 1% TX100 (▲), 0.5% Brij 96 (o), or no detergent (▽). Centrifugation was run at 40,000 rpm at 4°C for 19 h in a SW 41 rotor (Toutant and Massoulié 1987).

2. Low salt extracts of *Drosophila* heads with 1% TX100

AChE from *Drosophila* heads was extracted (w/v 1:10) in a low salt buffer containing 1% TX100 and a cocktail of antiproteolytic agents. Three identical samples were incubated with 0.5% TX100 (sample a), 0.5% Brij 96 (sample b), or an equivalent volume of buffer (sample c) at room temperature for 20 min and intermittently stirred. Samples were then centrifuged in 5-20% sucrose gradients containing 0.5% TX100 (a), 0.5% Brij 96 (b), or no detergent (c). Figure 2 shows that AChE consisted in a major G_2 form which sedimented at 6.1 S in a), at 4.2 S in b), and aggregated in the absence of detergent (c). The aggregation of the amphiphilic G_2 forms in the detergent-free gradients made it possible to detect two minor components corresponding to hydrophilic dimer and monomer.

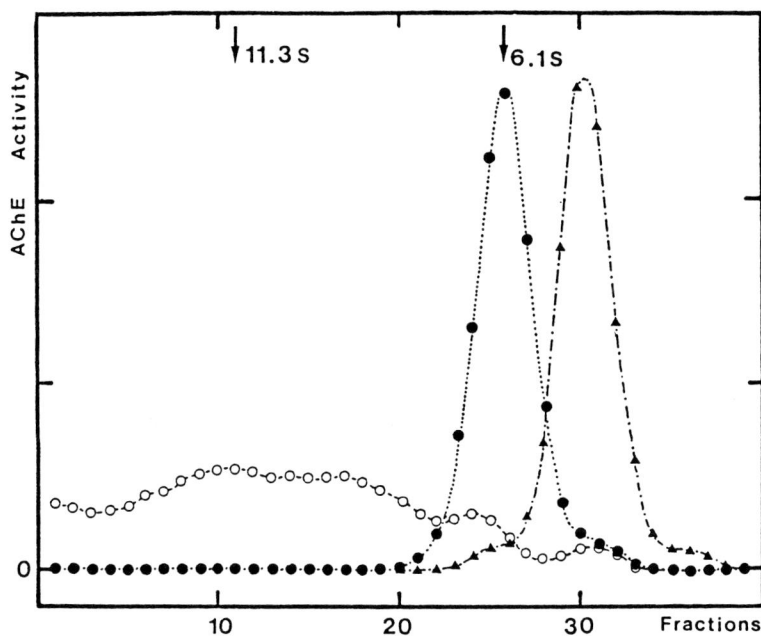

Fig. 2. Analysis of AChE molecular forms in a detergent-soluble fraction of *Drosophila* heads. Centrifugation in low salt 5-20% sucrose gradient containing a) 0.5% TX100 (●), b) 0.5% Brij 96 (▲), or c) no detergent (o), run at 40,000 rpm at 4°C for 18 h (Toutant et al. 1988).

Non-denaturing electrophoresis

Helenius and Simons (1977) introduced charge-shift electrophoresis, a simple and powerful method to distinguish between amphiphilic and hydrophilic proteins. The migration of amphiphilic proteins is accelerated in the presence of the negative charged detergent deoxycholate (DOC) added to the non-ionic detergent TX100 in the running buffer, and it is slowed down in the presence of the positive charged detergent cetyltrimethylammonium (CTAB).

We adapted this method to the study of AChE globular forms from chicken embryo leg muscles (Toutant 1986). We observed that the migrations of amphiphilic forms were influenced by ionic detergents, and we defined a hydrophobicity index (HI):

$$HI = \frac{m(DOC) - m(CTAB)}{m(TX100)}$$

in which m(DOC), m(CTAB), and m(TX100) are the migrations (in mm) of a given AChE component in the tree types of gels. HI is zero for hydrophilic molecules and has positive values for amphiphilic proteins which increase as a function of the amount of detergent bound. HI, therefore, quantifies the degree of interaction with detergents. This method showed that the G_1 and G_2 forms are amphiphilic and allowed the identification of three subsets of G_4 form in chicken embryo muscles (one hydrophilic component, and two amphiphilic components differing in the value of HI) (Toutant 1986).

Since CTAB bears a quaternary ammonium group and has an inhibitory effect on the AChE activity in certain species, this method may be simplified: we describe in the next section two applications on non-denaturing electrophoresis to *Drosophila* AChE (Toutant et al. 1988) and *Torpedo* BuChE (Bon et al. 1988, in press).

1. *Drosophila* AChE

A crude extract of *Drosophila* AChE (in low salt buffer containing 1% TX100) was auto-lyzed by storage at 4^oC. We observed that the major amphiphilic G_2 form (see Figure 2) was progressively converted into hydrophilic G_2 and G_1 forms.

We analyzed such an extract in non-denaturing electrophoresis. Gels (7.5% polyacryl-amide) and running buffer contained 5×10^{-2} M Tris-glycine, pH 8.9, and either 0.5% TX100, or 0.5% Brij 96, or 0.2% DOC + 0.2% TX100, or no detergent. The electrophoresis was run under 10 V/cm for 3 h. AChE activity was revealed according to Karnovsky and Roots (1964). Before loading the gels containing Brij 96 or DOC, samples were incubated with the running buffer (usually 1:1) at room temperature for 20 min in order to allow the exchange of TX100 present in the extract.

Figure 3A shows that three AChE components (1, 2, and 3 in the order of increasing mobilities) were detected in the gel containing 0.5% TX100: they correspond to an amphi-philic dimer (1), a hydrophilic dimer (2), and a hydrophilic monomer (3). In the presence of Brij 96 (Figure 3B), we noted that the migrations of components 2 and 3 were unchanged but that component 1 was slowed down. Expressing the migration of amphiphilic dimer as a per-centage of that of hydrophilic dimer, we found 33% in A and only 19% in B. In the absence of detergent in the gel (Figure 3C), a part of component 1 precipitated in the gel (as a thin line near the well), and a smear of activity was also observed (lane T, Figure 3C). We noted that a preparation of amphiphilic dimers recovered as aggregates from a preparative gra-dient run without detergent (lane 1, Figure 3C), did not enter the gel. Thus, the behaviour of amphiphilic molecules in a gel devoid of detergent depended on the amount of TX100 remaining in the sample. In the presence of 0.2% TX100 + 0.2% DOC (Figure 3D), the migration of component 1 was 80% of that of component 2.

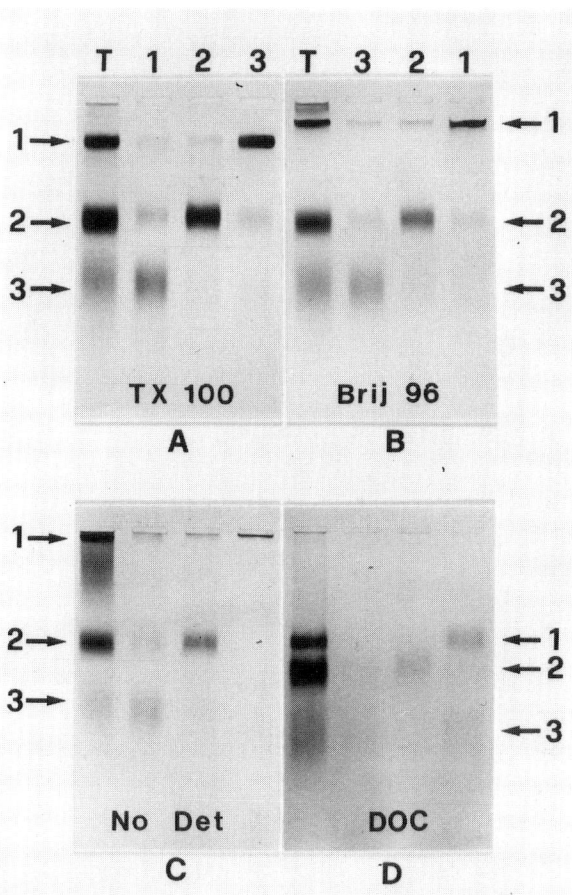

Fig. 3. Non-denaturing electrophoresis of *Drosophila* AChE. The three components present in the total DS fraction (lane T) were identified as amphiphilic dimers (1), hydrophilic dimers (2), and monomers (3) by electrophoresis of peak fractions isolated from preparative centrifugation (lanes 1, 2, and 3). 7.5% polyacrylamide gels and running buffer contained (A) 0.5% TX100, (B) 0.5% Brij 96, (C) no detergent, or (D) 0.2% DOC + 0.2% TX100. Migration at 10 V/cm for 3 h. (Toutant et al. in press).

2. *Torpedo* heart BuChE

We previously reported that BuChE was the only cholinesterase present in *Torpedo* heart (Toutant et al. 1985). The globular forms of the enzyme were extracted successively in a low salt buffer without detergent (low salt-soluble or LSS fraction), and then in a low salt buffer containing 1% TX100 (detergent-soluble or DS fraction). The analysis of LSS-BuChE and DS-BuChE by sucrose gradient centrifugation showed that LSS-BuChE possessed a non amphiphilic G_4 form and amphiphilic G_2 and G_1 forms. DS extracts contained amphiphilic G_4 BuChE and αG_2 AChE form representing an erythrocyte contaminant. We observed that amphiphilic forms of the DS extract aggregated in the absence of detergent in the gradient, whereas amphiphilic forms of the LSS fraction were non-aggregating.

The different globular forms of heart BuChE were analyzed by charge-shift electrophoresis: Figure 4 shows that this method constitutes a valuable tool in the characterization of amphiphilic and hydrophilic components. In addition, it allows the separation of molecular variants which are not distinguished by their sedimentation coefficients.

Electrophoresis was performed in 7.5% polyacrylamide gels containing no detergent (W, without detergent), 0.5% TX100 (T), or 0.5% TX100 + 0.2% DOC (D), run at 15 V/cm for 3 h. In each gel, a sample of globular lytic AChE derived by proteolysis from asymmetric forms of electric organ was loaded in lane 1, and was used as an internal marker of migration. In the other lanes, we compared total extracts (LSS, DS fractions, and a mixture of LSS + DS), and individual peaks recovered from preparative gradients ($G_2{}^a$ and $G_4{}^{na}$ of LSS-BuChE; $G_2{}^a$ and $G_4{}^a$ of DS-BuChE). The $G_1{}^a$ form of LSS-BuChE could not be isolated as it is readily converted into a $G_2{}^a$ form, especially in the presence of TX100.

The amphiphilic nature of G_1 and G_2 in LSS and DS fractions, and of the G_4 form in DS fraction was demonstrated by a significant acceleration of their migration in the presence of DOC. The G_4 form of LSS fraction had similar migrations in the three gels, and was thus clearly hydrophilic (non-amphiphilic). The DS G_2 amphiphilic form is recognized by an anti-AChE antiserum, and its electrophoretic migration corresponds to that of erythrocyte AChE.

COMMENTS

For several reasons, cholinesterases constitute an ideal model for the definition of operational criteria characterizing amphiphilic and hydrophilic (non-amphiphilic) proteins.

Firstly, these enzymes exist under a number of molecular forms including non-amphiphilic globular forms (G^{na}), and amphiphilic globular forms (G^a). As discussed below, the latter forms are known to possess at least two types of hydrophobic domains, and we distinguished two classes differing in their capacity to aggregate in the absence of detergent (Bon et al. 1988, in press). It is thus possible to analyze the behaviour of various types of amphiphilic forms, and to compare it with that of hydrophilic forms, either naturally occurring, or the

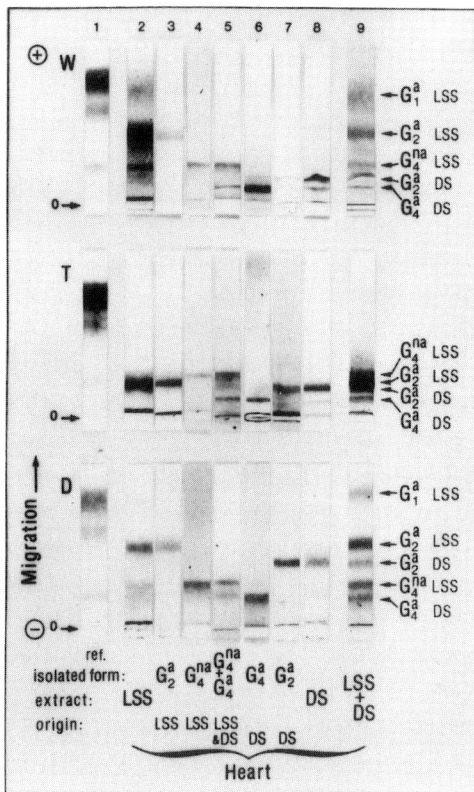

Fig. 4. Characterization of amphiphilic and non-amphiphilic components of heart BuChE by charge-shift electrophoresis. Total LSS and DS fractions and molecular forms isolated from sucrose gradients were analyzed in 7.5% polyacrylamide gels containing no detergent (W, top), 0.5% TX100 (T, middle), and 0.2% DOC + 0.5% TX100 (D, lower panel). The gels were stained with acetylthiocholine as a substrate with no inhibitor: Lane 1: lytic G_2 and G_4 forms of *Torpedo* electric organ AChE as a reference; lane 2: LSS fraction; lane 3: amphiphilic G_2 form of LSS fraction; lane 4: non-amphiphilic G_4 form of LSS fraction; lane 5: combination of LSS non-amphiphilic G_4 form with amphiphilic G_4 of DS fraction; lane 6: amphiphilic G_4 form of DS fraction; lane 7: amphiphilic G_2 form of DS fraction (erythrocyte AChE contaminant); lane 8: total DS fraction; lane 9: combination of LSS and DS fractions (Bon et al. 1988, in press).

lytic derivatives obtained by digestion of the asymmetric collagen-tailed forms (by proteolysis), or of the amphiphilic forms (by proteolysis, and in some cases by phosphatidylinositol phospholipase C (PI-PLC)). It is particularly useful to include such hydrophilic standards in non-denaturing electrophoresis.

Secondly, these enzymes may be detected in a simple and sensitive manner by their catalytic activity, either in solution, or in gels using a spectrophotometric assay or histochemical methods. This makes it possible to analyze extracts from a variety of tissues which may possess a low level of cholinesterase activity. It is important to avoid additional steps of concentration and purification during which modifications of the properties of the molecules may occur, for example by autolytic processes as described here in the case of *Drosophila* AChE.

Amphiphilic proteins have been defined by their capacity to associate with micelles of non-denaturing detergents. This binding may be demonstrated by changes in the macromolecular parameters of the proteins upon addition or removal of detergent, and by differences observed in the presence of various detergents. This is illustrated here by sedimentation and electrophoretic mobility. Other parameters such as Stokes radius, may also be used, but in a less convenient manner. The method of charge-shift electrophoresis takes advantage of the difference between neutral, positive and negative charged detergents. We observed that the two neutral detergents TX100 and Brij 96 (their chemical structure is shown in Figure 5), also show an interesting difference: Brij 96 slows down the sedimentation as well as the electrophoretic migration of cholinesterase amphiphilic forms more markedly than does TX100, possibly because of the larger size of its micelles. This property makes it easier in some cases to identify amphihilic forms, for example by increasing the difference in the sedimentation of two components in a tissue extract.

We observed that amphiphilic G_1 and G_2 forms of AChE and BuChE are often solubilized in the absence of detergent (i.e., in the LSS fraction). Detergent-interacting ChE forms can also be secreted (Lazar et al. 1984), and may occur in plasma (Bon et al. 1988, in press). Alternatively, hydrophilic forms sometimes require detergent to be solubilized (for example, a hydrophilic G_4 form in chicken embryo leg muscles (Toutant 1986)). These observations demonstrate that solubilization requirements cannot be used to characterize amphiphilic molecules. There is no equivalence between "soluble" and "particulate" fractions on the one hand, and hydrophilic and amphiphilic proteins on the other hand.

The structural organization of the hydrophobic domain of amphiphilic forms of AChE is now known in a few cases. Amphiphilic G_2 forms of human erythrocytes, *Torpedo* electric organs, and *Drosophila* heads possess a covalently attached glycolipid at their C-terminus. The G_4 form of AChE from bovine caudate nucleus possesses a non-catalytic hydrophobic subunit which is linked by disulfide bonds to one of the dimers (Inestrosa et al. 1987). Although these modes of membrane insertion are probably valid for G_2 and G_4 forms of cholinesterases in different species, it should be stressed that several observations indicate a certain heterogeneity in the amphiphilic properties even within a given species. For example, we recently observed that amphiphilic G_2 forms of *Torpedo* tissues can be divided into two classes (Bon et al. 1988, in press): class I forms aggregate in the absence of detergent, at least at high concentration, or in the presence of a detergent extract of electric organ, and they are sensitive to PI-PLC which converts them into hydrophilic derivatives; class II forms do not possess these properties. It should be noted that the lack of sensitivity to PI-PLC, proves by no means that these molecules necessarily differ in the nature of their hydrophobic domain-since the well characterized glycolipid-anchored G_2 AChE from human erythrocyte, and is itself PI-PLC resistant (Roberts et al. 1987).

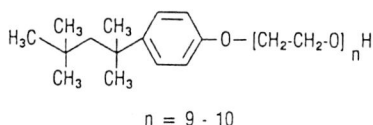

$$H_3C \cdots \begin{smallmatrix} CH_3 & CH_3 \\ | & | \\ | & | \\ CH_3 & CH_3 \end{smallmatrix} \cdots \text{—O—} [CH_2\text{-}CH_2\text{-}O]_n H$$

n = 9 - 10

Triton X100 = Polyoxyethylene(9-10)octylphenol

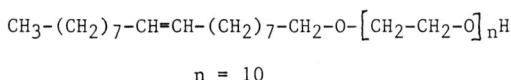

$$CH_3\text{-}(CH_2)_7\text{-}CH=CH\text{-}(CH_2)_7\text{-}CH_2\text{-}O\text{-}\left[CH_2\text{-}CH_2\text{-}O\right]_n H$$

n = 10

Brij 96 = Polyoxyethylene(10)oleylether

Fig. 5. Chemical structure of Triton X-100 and Brij 96. Both non-ionic detergents are of the polyoxyethylene type (similar hydrophilic heads), but differ in the extent of the hydrophobic tail (adapted from Helenius and Simons 1975). Brij 96 was purchased from Sigma (P 6136), and TX100 from Merck.

REFERENCES

Arpagaus M, Toutant JP (1985) Neurochem Int 7:793-804
Augustinsson KB (1963) In: Koelle GB (ed) Cholinesterases and Anti-Cholinesterase Agents. Springer Berlin Heidelberg, pp 89-128
Bon S, Massoulié J (1980) Proc Natl Acad Sci USA 77:4464-4468
Bon S, Toutant JP, Méflah K, Massoulié J (1988) J Neurochem (in press)
Borsum T, Bjerrum OJ (1986) Electrophoresis 7:197-203
Brodbeck U (1986) In: Watts/De Pont (eds) Progress in Protein-Lipid Interactions 2. Elsevier Science Publishers Amsterdam, pp 303-338
Fournier D, Cuany A, Bride JM, Bergé JB (1987) J Neurochem 49:1455-1461
Futerman AH, Low MG, Silman I (1983) Neurosci Lett 40:85-89

Futerman AH, Fiorini RM, Roth E, Michaelson DM, Low MG, Silman I (1984) In: Brzin M, Barnard EA, Sket D (eds) Cholinesterases. Fundamental and applied aspects. W. de Gruyter Berlin New York, pp 99-113

Gnagey Al, Forte M, Rosenberry TL (1987) J Biol Chem 262:13290-13298

Grassi J, Vigny M, Massoulié J (1982) J Neurochem 38:457-469

Helenius A, Simons K (1975) Biochim Biophys Acta 415:29-79

Helenius A, Simons K (1977) Proc Natl Acad Sci USA 74:529-532

Inestrosa NC, Roberts WL, Marshall T, Rosenberry TL (1987) J Biol Chem 262:4441-4445

Karnovsky MJ, Roots L (1964) J Histochem Cytochem 12:219-222

Lazar M, Salmeron E, Vigny M, Massoulié J (1984) J Biol Chem 259:3703-3713

Massoulié J, Bon S (1982) Ann Rev Neurosci 5:57-106

Massoulié J, Toutant JP (1988) In: Whittaker VP (ed) The cholinergic synapse. Handbook of Experimental Pharmacology. Springer Berlin Heidelberg (in press)

Melanson SW, Yun CH, Pezzementi M, Pezzementi L (1985) Comp Biochem Physiol 81C:87-96

Ott P, Brodbeck U (1984) Biochim Biophys Acta 775:71-76

Ott P (1985) Biochim Biophys Acta 822:375-392

Roberts WL, Kim BH, Rosenberry TL (1987) Proc Natl Acad Sci USA 84:7817-7821

Römer-Lüthi CR, Ott P, Brodbeck U (1980) Biochim Biophys Acta 601:123-133

Rosenberry TL, Scoggin DM (1984) J Biol Chem 259:5643-5652

Silver A (1974) The Biology of Cholinesterases. North Holland Amsterdam

Stieger S, Brodbeck U (1985) J Neurochem 44:48-56

Toutant JP, Massoulié J, Bon S (1985) J Neurochem 44:580-592

Toutant JP (1986) Neurochem Int 9:111-119

Toutant JP, Massoulié J (1987) In: Kenny AJ, Turner AJ (eds) Mammalian ectoenzymes. Elsevier Science Publishers Amsterdam, pp 289-328

Toutant JP, Arpagaus M, Fournier D (1988) J Neurochem 50:209-218

Subject Index